GOD'S
BRAIN

GOD'S
BRAIN

LIONEL TIGER
MICHAEL McGUIRE

 **Prometheus Books**

59 John Glenn Drive
Amherst, New York 14228–2119

Published 2010 by Prometheus Books

Inquiries should be addressed to
Prometheus Books
59 John Glenn Drive
Amherst, New York 14228–2119
VOICE: 716–691–0133
FAX: 716–691–0137
WWW.PROMETHEUSBOOKS.COM

14 13 12 11 10   5 4 3 2

Library of Congress Cataloging-in-Publication Data

Tiger, Lionel, 1937–
    God's brain / by Lionel Tiger and Michael McGuire.
        p.   cm.
    Includes bibliographical references and index.
    ISBN 978–1–61614–164–6 (cloth : alk. paper)
    1. Psychology, Religious. 2. Religion. 3. God. 4. Brain. I. McGuire,
Michael T., 1929– II. Title.

BL53.T54 2010
200.1'9—dc22

                                                                    2009047153

Printed in the United States of America on acid-free paper

CONTENTS

PREFACE

The first impulse animating this book was simple if enchanted puzzlement about the remarkable difference between what the brain created about religion and the vast and long-lasting social systems that were the result. This is obviously an extraordinarily important aspect of human behavior that has to be understood as skillfully as possible. But we were troubled because so much of the public dialogue on the matter was beset by acidulous hostility from those opposed to religions on one loud and clamorous side. On the other, there was self-confident certainty of both people and even governments about the need to quarantine the modes of faith from questions and from doubt.

This has ceased being a theoretical matter. Its episodically fierce practical implication has leveraged sharply our original scientific focus. We reside in a country in which religious belief is an active feature of government and, in a tacit sense, a prerequisite for elected public responsibility. As well, the United States is inextricably embedded in a world in which the fires of

7

faith stoke entire powerful governments. Despite the usual overflowing basket of human needs, sorrows, failures, and traumas therein, there is in countless communities an overlay of faith-stoked enthusiasm and often-militant ardor to achieve the perfect triumph of one system of belief or another. People are killed for what is in their heads and for the amulets worn around their necks. A mighty university press, Yale, refuses to publish images of a religious figure—the book is precisely about them—because of fear that grandly confident and irritated zealots will turn anger into the murder of editors, as they already have. The quiet solace of gentle, lawful community to be found in local places of customary religion exists alongside terrifying violence that asserts the superior value of one style of solace compared with another. Amid the jangle of international politics and search for group significance, religion is often the herd of elephants in the crowded conference room. Residual respect for what was once the calm core of communal order embargoes any talk of what impulse or need or artfulness underlies it all. And the human brain remains hard at work.

Michael McGuire
Lionel Tiger

ACKNOWLEDGMENTS

Writing acknowledgments is one of the pleasant parts of writing books in that it allows appreciation and credits to be directed where they belong.

Robin Fox, Chris Boehm, John Chandler, and Nancy Brown read early versions of the book and provided helpful comments and support. Jay Feierman, who read several versions of the manuscript, and Mark Hall provided detailed critiques and many suggestions for change that improved the quality of the book significantly. Throughout the development of this book, Beverley Slopen has been a persevering and extremely encouraging agent, and she receives our special thanks.

Then there are those who through discussion have influenced the content of the book. These include Monica Gruter Cheney; Roger Masters; Stuart McGuire; members of the Gruter Institute and the International Society for Human Ethology, where at their meetings presentations of some of the material in the book were made; and, finally, several individuals who have shared personal information but who have requested they not be named.

Chapter 1

AND WHAT AN AMAZING, IF IMPROBABLE, STORY IT IS

Dazzling Spanish cathedrals. New England steepled chapels. Sunday-service slum storefronts. Vatican City. Shinto shrines in Kyoto. These are physical expressions of the 4,200 reported, distinct faith groups—another term for religion—in the world.[1] They reflect why 370,000,000 Google citations are provided a curious, hapless researcher who types in *religion*. Religion as a process generates remarkable action, countless events, and numberless provocative artifacts. Yet what factual phenomenon except perhaps slips of ancient holy paper underlies and animates one of the most influential and durable of human endeavors?

We've an answer.

Shivers in the moist tissue of the brain confect cathedrals.

Our proposal is that all religions differ but all share two destinies: they are the product of the human brain. They endure because of the strong influence of the product—religion—on brain function. The brain is a sturdy organ with common characteristics everywhere. A neurosurgeon can work confidently on a Vatican patient and another in Mecca.

Same tissue, same mechanisms. And one such mechanism is a readiness to generate religions. Another is to respond positively and with conviction to what it has generated.

One brain, four thousand religions.

WHERE TO START?

This proposal may appear at once too simple and too broad. And it is. But religion has been and is so drastically important to human beings that it requires firm and compassionate analysis. We have to start somewhere to assemble such an analysis. Religion does not derive from the elbow or the liver but from the remarkable managerial organ that is the common factor to all beliefs.

This is where we begin. Time for religious brain surgery.

An initial surprise is the inexorability and roundness to the story of religion that reflects it creates a problem—of explanation and direction of the most difficult issues of life—and then sets about solving it. It's a self-supporting, self-reinforcing, self-solving system. Remarkable. The ever-active human brain perceives or fears or understands some drastic factors in the world, such as death and evil and uncertainty and failed love, which have to be explained. Then the brain sets about answering the questions raised. Do you remember the childhood toy featuring a bird with a beak poised on the lip of a glass of water that dipped into the water and then returned to its position and then dipped into the water and then returned, and so on?

This circularity in religion is at once simple and puzzling: religions create explanatory problems and then produce and determine the conditions of their remedy. It is a striking and intricate product of the brain. It is at the same time embedded in a physiology with cycles, illness, fatigue, excitement—this is the equivalent of a weather system within which the brain must operate.

The body is the brain's environment. Religious observances and beliefs appear to discipline or at least soothe both large (dying) and small (a controlled flirtation) dramas of the mortal coil.

Our effort here will be to show how the connections between body, brain, and social group sustain, threaten, or isolate the religious process.

We are convinced the matter is so fundamental that it is more than likely that the broad impulse served us for thousands of generations. It comes with the sapient *Homo* territory. The specific expressions of the impulse vary enormously —Buddhist, Islamist, animist, whatever—but the ongoing process varies far less. Meanwhile, the consistencies lure us.

THE NATURALIST IMPERATIVE

Explaining religion and belief requires permissive imagination and breadth of analysis. Religious belief stimulates everything from sweet spring holidays of solemn contemplation in Kyoto to Crusades and lethal jihadist wars. It helps fill athletic arenas in Texas with admirers of the gripping theatrical messages of brilliant preachers who declaim narratives they insist have divine authorship. Their virtuosity is in thrilling rendition, not creation. It urges persons into confessional cubicles where in secret to all but the Almighty they reveal their private fears, failures, suspicions of self, and caution about motives. It impels them in Arabia to an annual parade during which they whip themselves with sharp chains. It hosts Quakers who seat themselves in quiet group intensity in elegantly austere chapels to reflect on meaning and the colors of life. It supplies a religious figure to accompany the execution of persons a community decides are loathsome. Understanding such behavior demands virtually limitless patience and disciplined suspension of censure.

There has been a spate of books about religion that have been very popular. But the most prominent and influential specialize in censure, not analysis without rancor. The most prominent authors are Richard Dawkins, Christopher Hitchens, and Sam Harris.[2] While we may agree or disagree with their various diagnoses of the social impact of religious behavior, we are principally concerned with their point of departure. A gentler and more measured review asserts the primacy of science while acknowledging the historic attractiveness of religious explanations surrounding issues of life's foundation.[3] Our only task is evocation and analysis.

There are vast differences among religious groups. They knowingly and eagerly define themselves by their special rules of behavior, their dress, their sacred days and places, and by what they in particular define and even cherish as unknown. And the unknown is remarkably durable. For many people it's almost more tangible than anything else. People may change their enthusiasm for authors or brands of yogurt or piano players or even political parties. But they are less likely to alter their stance on matters involving religion—either to take one up or let one down. This may achieve improbable comedy at times. In an interview with Terry Gross on National Public Radio in the United States, the comedienne Carol Leifer, who had been married to a non-Jewish man and then divorced, described her traditional parents' happy and relieved response when she moved in with a Jewish woman. Leifer noted, "They would have welcomed a chimp so long as she was Jewish."

Legion and admirable explanations exist for these turbulent matters. Initially, most efforts at explanation were broadly theological. So it was no accident that the earliest and greatest universities began way back when as religious institutions. Religious groups owned the human narrative and could control the formal languages of interaction too, as with Latin— the unstained Microsoft Word of the old days. But now, of course, there has been immense change in the portfolio of explanations offered of religion just as universities are forever changed from centers of spiritual contemplation and thought.

The new explanations have been and are rooted in psychological, cognitive, historical, evolutionary, and related forms of analysis. Many distinguish between the various flavors of belief and religion. They may sustain enthusiasm for one flavor and disdain for another or all others. They identify their features and locate them in both historical and contemporary events. Others postulate a "God gene" or some equivalent, which somehow generates religious dogma and behavior. Still others advance the idea that "God put religion in the human brain." And yet others focus on the possible adaptive value of religion should it be shown that religiosity moderately increases health, survival, and reproduction (it appears to do so often enough to compel attention). The list of explanations is long—with no end in sight. And those who support this variable array of certainties are some of the most tenacious advocates it is possible to imagine.

We need to learn something new about something old and omnipresent. We need precise knowledge about something as general as belief. New, good information about many cultures helps us. Provocative science about the brain and its innards tempts us. Virtually all human groups generate and sustain religions.

It is simply good naturalistic science to begin and not end with that huge reality.

THE SOURCE?

Gravity holds down the cathedral's stones but what builds them up? God? Belief or faith? Assertive books? Revelations? In fact, yes. These are experienced and retold and certified as the grand cause of remarkable and epic outcomes. But what caused these causes is never announced. They are like air. They just *are*.

But the source, the source! *They do not satisfactorily explain why faint electrical and biochemical processes located principally in the human brain can generate the remarkable and durable human processes that concern us here.*[4] The belief-something in the brain, that old belief-something, keeps ticking consistently. It is sustained independently of and even despite graphic empirical evidence. Perhaps the world is supposed to end next October 18 but doesn't. Rest assured that the Eighteenth-October-Day Adventists with the appropriate conviction will swiftly generate a new explanation for the change. The new holy plan will quickly satisfy those who accept it.[5]

Perhaps earthquakes and wars and incomprehensible disasters occur. Nonetheless some faith in a benign overall operating system remains in place and may even find its impact emphasized in adversity. Countless commentators insist on repeating the warning, "Imagine if there were no faith!" Their audience appears to be largely appreciative.

Pangloss and Voltaire are still in the building. The focus is everywhere but on the source.

AN ARENA WITHOUT BOUNDARIES

Groups knowingly and often proudly define themselves by their creativity, such as their ability to concoct gods, rituals, and beliefs. These may be incomprehensible in terms of logic and impossible to prove in terms of science, though powerful forces seek to at least blur the boundaries between science and religion to the advantage of neither and no one. Such stubbornness has produced a conflict that has historically and at this troubled instant preoccupied those who puzzle about the sources and nature of religion—between those who believe and those who scorn believers.

It's some context! Believers firmly scorn the austerity of secularists. Nonbelievers are thought to build and use dull buildings in which they play poor music, perform prosaic celebrations in which officials sport no special or colorful costumes. And where is the joy in the ascetic and banal descriptions of nonbelief? Why is nonbelief so dull? For their part, nonbelievers may scorn what they see as the frail scantiness and improbability of grand beliefs—even if they're celebrated in elegant choir-rich cathedrals. They parse the arguments, trace the evidence, and come away, their heads shaking with some bewilderment.

And, not surprisingly, disagreements abound. Endless arguments about evolution, intelligent design, atheism, and the benign impact of belief on morals and comportment have consumed time and energy, before Galileo and since.

Whatever view of these matters one adopts, it remains clear that religions can and often do exert a remarkable grip and influence on what people think and how they behave. For some, they invite a lifetime of devotion—for example, St. Francis of Assisi and Mother Teresa. For others, they motivate action that can achieve wild fatal efficacy—religions can convince believers it is imperative or at least desirable to kill people who believe or behave differently than they do. Death to . . . whomever!

Death to nonbelievers? Can you believe this? Why not just ignore those wayward infidels and spare yourself the work, irritation, and mess of killing them? The Taliban's work is never done.[6]

Explanations, please. Christians, Jews, and Islamists evoke the certified authority of their gods, texts, rituals, and rules of behavior. They define a comprehensive list of sins and virtues.

They provide often-strict accountancy for the prospect of an afterlife marked by a variety of conditions. Confucians emphasize their ethical precepts. Taoists tie personal conduct to how the universe functions. Groups such as Mormons offer detailed programs for personal and communal behavior, which are derived from relatively recent revelations.[7]

And—please, please—savor how extraordinary this is: most religions promise an afterlife of varying degrees of reward and punishment, flame and freshness, pleasure and pain. And it is simply impossible to overestimate the drastic importance of this promise. There is nothing with which to compare it. What a person's dance card will reveal in death is never clear to the living. But the entire matter is deeply involving. Recall that for centuries Catholics purchased indulgences, with real money now, to enhance conditions in the next world. To this day, Americans donate more money to religion than to any other sector of their society. Belief in another life later affects directly the daily choices believers make in all of life's important realms. It's a spectacularly attractive idea that while living stops, life doesn't. If religion as a force has one unique and signature product, it is the afterlife.

And so religious beliefs are coveted, often tenaciously, when they proffer the lure of a potentially wonderful afterlife. We will discuss shortly how and why they have a comforting impact on brain and body physiology—the core of our book— and why different religions differ in efficacy. Just as some foods produce calories and energy more agreeably and efficiently than others, religions perform their work with varying degrees of success. Religious affiliations usually endure throughout the life cycle. For six or seven decades a person will feel linked to a concept of being, its grand origin and history, and its expression in daily life. Important actions such as marriage and parenting follow its rules. Minor details (which may sometimes assume massive importance), such as which foods to eat and reject and which books or films to consult, may be affected by a web of obligation rooted in an ancient and distant story.

WHERE TO LOOK, WHERE TO LOOK?

Have those seeking explanations for this been working in the wrong places with inadequate tools? Here is where we will look and why.

Nature depends on redundancy and a kind of functional overkill. It takes little for granted and prepares for the worst. Most of our systems can do much more and much better than they usually do. Athletes prove this, though most, of course, must exert themselves to make the point. The brain is no exception. It has the capacity to imagine decisions far more important and dangerous than which toothpaste to choose and which people with whom to associate during a quick lunch or the long life cycle. And that takes discipline too—such as the theological universities of seventeenth-century Europe in which there was endless and endlessly sophisticated disquisition on the nature and form of faith. And of course this was the bedrock character on which much of the North American higher educational system was built as well, into the early twentieth century.

Even when many of the banal and repetitively tiresome details of living are handled satisfactorily, or the brain has preoccupied itself with theological questions and facts it can't answer, the brain's owner will ratchet up the action. He or she will seek a thriller film or book or hike or ski alarmingly quickly or ride a large horse or undertake an immense array of other inessential but satisfying activities. The brain needs a kick and a change. Like that toy bird on the lip of the glass of water, it needs action, movement, a system.

And the change the brain seeks is to *brainsoothe* itself, to reduce its preoccupations and acquire the most interesting menu of stimuli it can safely find. It craves practicing the tool kit of socioemotional capacities with which it has been endowed by evolution and with which demanding daily events appear to be in frequent or even constant opposition.

Religion is to the brain what jogging is to the legs. It is a form of socioemotional and institutional exercise for the organ in our head.

And the products of this exercise are clear: just as trained legs aid in producing sprinters, hurdlers, strollers, challengers of Everest, and Baryshnikov, the human brain produces the vast array of beliefs, superstitions, astrologers, religions, and armies of the faithful. It is directly responsible for Scientology, the pope, Iowan missionaries in Kenya, the anaesthetizing Book of Mormon, and canvas-vested bombers bent on suicide to soothe Allah. But it also helps produce the still and affable center of countless communities and parishes and temple neighborhoods. In many places, it is the boss of birth and marriage, where there is no distance between living fairly and with civility and sustaining a religious belief and practice, which roots the individual and dignifies the social group.

Therefore, we cannot as a point of departure divide our subject matter into good guys and bad guys, fine ladies and mean ones. And we won't, if for no other reason than that they all have examples of the human brain in their skulls. This cannot mean that all behavior is equally sound, equally fair, equally glorious, equally sensible, and equally good to have around. It only means we cannot begin with our conclusions.

Chapter 2

YOU NEED BOTH A ZOOM LENS AND A MICROSCOPE TO SEE RELIGION

When the memberships of all the Christian, Islamic, Hindu, Buddhist, Taoist, Confucian, Shinto believers, and so on are tallied up, and religions based on special powers of animals or mystical symbols are added to these figures, well over 80 percent of the world's adult population is estimated to live their lives within the embrace of a known religion.[1] Religion is virtually everywhere as both a shadow and a light. The fact of religion is in the heart, a pew, an exploding suicide bomber, a holiday dinner for the poor, and in the heavens. Understanding it is difficult. The answer to the puzzle—Why? How? When?—is far from obvious. Small wonder it's always been a challenging issue both as a practical matter and an intellectual one.

But a beginning is necessary. Simple numbers—a census of the existence of religions—offer a generous, direct, and informative perspective. And yet, of course, the scale of the behavior involved is altogether enormous.

NUMBERS

Descriptive numbers are capacious enough—they go on forever, after all, from here to infinity—to accommodate the
range and reach of religion, and they can seem evenhanded
and fair. However, individual religions may seem odd and even
bizarre to some or many outsiders, and their benefits and
activities are hardly consistent from place to place and time to
time. But their overwhelming numbers and rich ubiquity
underline their normality in practice. We're dealing with a phenomenon as diffuse as oxygen and seemingly as imperative.

In the real social world, there are a reported 4,200 different
faith groups/religions.[2] This number is likely far short of the
true count. Currently the Bible is translated into 4,516 of the
6,912 languages spoken in the world.[3] There are numerous
translations under way into previously untargeted dialects—
it's difficult to know how many. There are a reported
2,100,000,000 Christians on the planet, 1,500,000,000 Muslims, and billions more in various faiths. And as we mentioned
earlier, type the word *religion* into the Google search engine as
we did on June 1, 2009, and there are 370,000,000 citations,
which could consume much of a lifetime simply to scroll
through—there will be thousands more next week, if not
tomorrow. No known country is without a religion or religions. We cannot avoid surmising that we are confronting a
phenomenon as ambient as gravity.

The numbers suggest the magnitude of the story but not
necessarily the elegance, banality, and forcefulness of religions.
They are an integral part of life, day in and day out. Not just
on Sundays, or Saturdays, or sacred days, or at sacred locations, or in this country but not that one. They are everywhere,
every day. They permeate. Countless people voluntarily stream
into church. Many read sacred texts or sermons or books on
religious subjects. Prisoners in some American institutions are

drilled in religious lore by spiritual organizations funded by taxpayer money.[4] Religious or not, lives are affected.

Numbers also can mislead. For example, actual religiosity in the behavioral sense may exceed recorded church attendance, as appears to be the case in a number of countries.[5] Estimates of committed individuals may be low. But polls can be tricky and influenced by the wording of questions, as, for example, in an Estonian poll during 2004 in which 49 percent of people surveyed said they didn't believe in God but, in weird hilarity, only 11 percent of those polled said they were atheists.[6] And of course, true or not, people may claim a religious affiliation, often the polite, conventional, or at least the easy thing to do, and almost obligatory among public officials in countless communities. For many pressed and busy people, a religious service may be an especially protected and relaxed period of the week or, more often, a certified time for "Don't bother me." And it is, of course, an important sign of the power of religion that during dire civil crisis, such as the Rwanda massacres, church buildings were often seen as protected havens. In gentler times, they are seen as safe or at least neutral spaces, relatively exempt from usual rules about trespassing and access.

The numbers describing religious phenomena are impressive. And both atheists and believers would not deny that religions deploy special energies. They offer something that humans appear to seek and have sought for all of written history and no doubt long before writing in the chill, moist, painted caves of the Dordogne and elsewhere where our ancestors gathered to depict animals, leave a palm-print, and scratch and paint stone with colored fingers. Religions breed ideas, reverence, fear, and mystery.[7] They demand commitment, compassion, sacrifice, subordination, resources, time, and work. They have irreplaceably catalyzed and supported virtually all the human arts. They have stimulated science and arts that have then turned against their original Catholicism of the Middle Ages.[8] They describe a soothing permanent eternity. With equal verve they weave sultry descriptions of hell that will never disappear. They are able to unsettle doubters for whom the deathbed conversion becomes an unexpected point of departure.

Now to belief.

BELIEF—A FIRST GLANCE

The beliefs of religions are as different as they are special. They are as diffuse as they are specific. They are at times simple and at times exceedingly complex. Their influence is everywhere— even in unexpected places.

Take as an example the beguiling aura of the perfect past— of Eden and paradise lost. It suffuses the present and the future. And surely this is truly an amazing and influential idea. Or, as Billy Graham once put it, *"It is impossible for those of us who have been created for eternity to find anything in the world to satisfy our souls fully."*[9] But what is the evidence for that paradise? What are its physical mementos? If there is no observable paradise or Eden on earth—by definition, of course—with what and on the basis of which evidence is the picture drawn of such assured and reassuring perfection? Yet it is the *very improbability* of such a perfect state when united with functioning religious activity that appears to render paradise plausible.[10]

This is pure belief. It is its own authority.

Belief becomes intensely personalized too, because, for example, finally and extravagantly an emissary from that past—a messiah or a prophet—will one day arrive, restore perfection, and radiate grace. But until then—no small matter— all involved must obey the rules and those who announce and administer them. Only obedient behavior today will ensure enjoyment of the stream of perfection that will be the holy future. The drama of all this can make the head spin. But it remains the week-in, week-out content of numberless churches and shrines and their members.

And, almost like no other, the power of religion's beliefs and their communication can be spellbinding, even immobilizing. As a child of eight, one of us attended an evangelistic church service with his grandparents in a small mining town in the Mohave Desert, renown for its overabundance of rattle and coral snakes as well as the fire and brimstone sermons of its preacher. The church was small, seating perhaps a hundred in wooden pews, and divided by a central aisle. The sermon was in full volume. The preacher pounded the pulpit as he described the horrors of hell. The audience almost dared not breathe or move—such goose bumps! Then, from under the front pew, a large rattlesnake emerged and slowly slithered down the aisle toward the exit (which suggests perhaps that religion is not located in that part of the brain humans share with reptiles— the rattler paid the preacher no heed!). There were whispers. Eyes turned to the snake. No one moved. No one spoke except the preacher. The sermon raged on. Finally, "A rattler!" a child screamed. The believers exited as one, with the preacher still preaching, "He won't bite, he won't bite, we're here in God's house." The church as haven, but still, a rattlesnake.

HIERARCHY

Human beings have been forever tortured by the contrast between the dream of equality and the regular recurrence of endless, invidious differences between people.

We want to show that religions are smack in the center of that nonstop argument. One of the main tasks they perform in the body social is to confront the fact of inequality.

Everyone may be equal in the eyes of God. But why are his eyes always looking downward? Why do ball players point upward to signal a homer when the earth on which they stand is so clearly implicated in their work? Why do the hands of prayer point to the ceiling or the vaulted dome? The ubiquity of human concern with hierarchy and social order suggests that the ability of religious impulse to confront and channel this concern is a feature of its efficacy and a not-so-secret secret of its success. We have to attend to this issue because it is intrinsically interesting. As well, it is revealing and significant that religions directly confront the provocations of equity and its opposite. And this must tell us something about the source of its potency. (Of course, Karl Marx famously thought that religion was a means for elites to control "the masses," which was a partial truth but also a large oversimplification. Elites often believe in much the same way as poor folk, however elegant are the surroundings within which their observances occur.)[11]

The provocative likelihood is that religions employ the subordination of men to gods as a mechanism for modeling subordination and dominance in society as a whole.[12] In this sense, Marx was on target. In effect, there is strong evidence from a host of potent sources that humans compose a species sharply marked by concerns about hierarchy.[13] The dream of equity is as powerful and recurrent as it has been because it is an antidote to the natural state of things—hierarchy.[14] Marx intuited that religious masks of inequality had to be abandoned for the triumph of classless society to emerge.

As a major social institution, a successful and durable religion has no option but to acknowledge the facts of the species with which it has to deal. It does so in varying degrees, from the polygamous cultish tyrants whose word is God and is derived from God to far more expansive and heterogeneous organizations, such as the Roman Catholic Church.

There is also one other all-pervasive reflection of how human beings express their hierarchical propensity to see the world in terms of up and down. This is contained in the general fact that, typically, members of each religious group decide that their particular flavor of belief and worship is best.

Who boasts that his or her religion is inferior to others? No one.

Nonmembers are presumed to be globally inferior or at least subordinate. We can perceive this in the famous Judaic claim that Jews are "God's chosen people" and the Islamic assertion of the intrinsic foulness of infidels. It is reflected in the fundamental faith in the triumph of salvation as a result of the drastic event of conversion. A new member of particular Christian groups may even be considered born-again, in an interesting version of the more Eastern notion of reincarnation, which here takes place within only one life cycle.

There seems little question that whatever the factual veracity of such proposals may be, the individuals who belong to vaunted groups are able to *enjoy* the experience of dominance over nonbelievers. This dominance becomes an odd and abstract, but emotionally real, benefit of membership. It is one of the most general but also most ubiquitous forms of asserting dominance available to ordinary people. Again, this is one of those features of religion the importance of which is impossible to overestimate.

One goal in this book is to untangle the threads that bind believers at once to their own displeasing frailty and to the fair and happy concert of heavenly grace. One question is: can a mammalian primate formidably equipped and even condemned to using the vocabulary of hierarchy create a language of glad connection between equals?

Like any two human beings, no two religions are quite the same. Each has its own identity, signature, and unique interpretation of sacred texts and history. Each has its own thought structure that seldom speaks with other structures. Each has its dogma and its rules about behavior. So at one temple women must cover their heads, while across the street men must. Seemingly arbitrary requirements of belief and action are required of members in good standing.

And the atmosphere of hierarchy predominates on both sides of the street. Each faith incorporates dramatically sculpted teachings of its founder. Some create the ubiquitous statuary of dominating religious figures. These memorable personages all, such as Jesus, Muhammad, Buddha, and Confucius are the foci of entire national libraries. They incorporate many of the ideas and values of the historical times in which they were founded. Their legends and authority bolster the impact of even ancient creeds on the immediate present with chronological alchemy.

We have identified two threads common to all religions. One is belief. The other is hierarchy—some structure of organization. To understand these, the first calls for theologians and neuroscientists, the second for sociologists and historians. Both are central to the matter. They depend upon each other. It is necessary to be aware of the different gradations of color in sacred costume and what their wearers mean to the ideas of religion's meaning. It appears that there cannot be a religious social order without someone issuing and accepting orders of one kind or another, from the most celebrated pope to the humble rural curate.

Orders, of course, may or may not be obeyed—religions have often been battlegrounds as well as havens. Or the connection between belief and hierarchy may be tacit and reluctant. There may even be a hierarchy established to prevent the formation of hierarchies. Nonetheless, they appear to be inseparable and play out their interdependence in often curious ways. But always hierarchy remains. The obligation of commitment to the meek and the poor in early Christianity is a poignant and impressive case in point. The Mother Teresas and Albert Schweitzers of the world vividly illustrate the difficulty of securing equity even among the most willfully devoted of hierarchical primates.

ACTION BRAIN—ACTION MAN

Whatever else members of religions do, they believe. This takes us directly to the brain. It's the brain that believes. Not the liver, the kneecap, or the elbow. We can now see the brain light up in the act of believing when people in a brain-scan machine are induced to contemplate their credos. We can measure changes in the brain's chemical makeup at such moments.

Belief is an organic product. It manifests itself not only in words or bowing or swaying or singing or participating. It reflects an organic, bodily relationship, almost as direct as the link between sweat and skin. The traditional and customary effort to create and sustain a distinction between body and soul is grandly, dangerously wrong. People of faith may object to this view. But from our perspective it will be impossible to appreciate fully the complexity and power of religion unless we move on from and abandon this false segregation of spirit from its instrument, the body. We will return to this idea.

The brain has evolved to identify and solve tasks essential for survival. *The brain evolved to act.* It evolved to *act*, not necessarily to think.[15] This is another central religious reality, the importance of which cannot be overestimated. The brain absorbs information and creates or imagines still more of what it considers information and combines the two. It organizes information for itself and is willing and often eager to explain to others what it has created. The brain along with its owner seem strongly inclined to believe what it explains. Together they predict events and initiate action and inaction.

There is a cascade of scientific reports about definable segments of the brain, what they do, and the details of how they work. This provides ever-more evidence about the physical brain's work in governing social behavior. For example, literally every week we learn more about which areas of the brain participate in identifying and responding to others.[16]

Can you believe that neuroscientists may have discovered pieces of brain tissue that made utilitarianism possible, that the greatest good for the greatest number can be found stamped on some collection of neurons?

The brain does these things naturally. It does so first during the earliest moments of infancy without parental training or the influence of peers. Even in utero it is sorting out different voices and motions and responds differently to different stimuli. A fetus will hear its mother preparing coffee and playing Mozart on the radio. Newborns track their mothers with their eyes, as if they were idolizing starlets. Such industriousness continues over the course of a lifetime.

It's easy to imagine why the brain might have evolved the way it has. It was adaptive because it had to be. Imagining that a predator may be hiding behind every large bush and taking appropriate precautions even without seeing one would save numerous lives. Imagining that there were other lands over the horizon might have led to exploration and discovery of more sympathetic or productive, or at least more interesting, environments. There are virtually countless possibilities.

And why might the brain work this way? It has to. The brain has to make decisions—to decide between real alternatives—to act, to survive, to reproduce, to prepare for difficult times. It has to learn to store food and wood for the winter. It has to learn to avoid mistakes or at least avoid repeating them, such as trying to navigate a swift river turbulent with springtime rains. It has to supervise going to the store before a national holiday. It has to fuss about why not to invite the boss for dinner. People like to and have to accomplish what's important to them, and the brain supervises how they transverse their environment. It does so because explanations and beliefs may improve control over events and increase the odds of predictable outcomes.

People want and seek control and predictability far more earnestly than uncertainty or failure. No one seeks failure and confusion. The pit-of-the-stomach feeling that gravity is abandoning us reveals the implacable connection between the brain's successful work, its failure, and the body's dependency on the outcome.

The brain can't avoid doing these things. Its job is to explain, to assess, to believe in its assessments, and to try to govern action. Again, the brain evolved first to act, and only then to think. Its job of survival is evolution's legacy, our intricate inheritance.

We've alluded to storing firewood, courting a mate, reassuring an infant, and fording a river. But the brain also imagines and believes things for which there is no *hard* evidence. What else could produce such astonishing ideas as the existence of life in other galaxies, gods, a designer of life on earth, animals with human motivations and personalities, an afterlife, hell, heaven, witches, demons, angels, and the certified sin of pride?

A word about hard evidence. It's a term scientists use. Most people don't reason as scientists do when they strive, systematically, to place limits on their own imaginings and beliefs. Scientists do this by distinguishing between ideas that can be disproved using scientific methods about hard evidence, and ideas that can't be disproved because there is no method for testing them.[17] Parapsychology is an example.[18] It is the view of most scientists that religious beliefs can't be tested. "Show me the science," as philosopher Daniel Dennett puts it.[19] Hence, beliefs are frequently characterized as imaginative but unsupportable.

This has been a central theme of critiques of religion by authors such as Richard Dawkins and Sam Harris.[20] There have been billions of words written supporting the ingenious proofs of potent deities developed by some of religion's most venerated thinkers, such as Saint Thomas Aquinas. There has been endless colloquy about reports of visions and revelations as with Joan of Arc, the Shroud of Turin, and the Burning Bush. But there is simply no scientific method to test for heaven; hell; thoughtful or considerate gods; and leering, masked, tail-flailing devils.

This assessment may seem harsh and, in a kind of heartbreaking way, it is. But that is not its intent. It's just the unavoidable outcome of scientific reasoning. The conflict between science and religion—the conflict between naturalism and theism—is not a game show.

But do these critiques of religion take into account the way the brain works? It is all well and good to argue that the brain should work a certain way—say, like the brains of scientists—and that beliefs should be buttressed by hard evidence. But if the brain doesn't work that way naturally, the argument is more wishful than real. Moreover, if religious belief is censoriously dismissed as "delusional" as some critics have done, it may lead to irritation or the fruitless silence of otherwise decent and thoughtful people. It will direct balanced inquiry away from religion, its attractions, its influence, its source, and most important, the brain at the heart of it all. And please don't be duped by the relativist's escape clause that all beliefs are equal. Far, far more is involved. There is no moral or conceptual equivalence here. A pound is not an ounce.

There is far more complexity to these points than appears at first. What most people's brains really do is organize information and make decisions about actions. They use a combination of what they know from their experience of the tangible world—what they see, smell, taste, hear, and touch—and from what they imagine. What they imagine are thoughts, ideas, scenarios, and explanations that they can't see, smell, taste, hear, or touch.[21]

Astonishingly, the human record shows that a bewildering portion of *what humans imagine is given the same weight, value, and authority as what they tangibly experience.*

The brain is simply more comfortable believing than doubting, just as the body is more comfortable lying in a hammock than hanging from chin-up bars. Consider how humans behave and make decisions. What is imagined as a guide to behavior is extraordinary, if not almost amazing. An employee imagines that a new job will be more fulfilling and satisfying than a current one, so he changes jobs. People go to a movie because they imagine they will be entertained, but it may or may not happen. They cook tacos for dinner because they imagine tacos will satisfy their hunger for taste and craft, but it may not happen. It's a plan, not a certainty. They buy tickets to a concert of new music, hoping to enjoy it. Or they believe that being nice to neighbors will improve their shot of getting into heaven, so they are nicer than they might wish to be. Or they believe that when they retire their golf handicap will improve.

These are all completely plausible choices with direct and tangible impact on people's lives. But during the same day they decide that if they say prayers once or twice or five times, they enhance their chances of going to heaven. In both wildly different cases, the brain arrives at a predictive conclusion. Once the brain imagines something, it seemingly can't let go easily without developing an explanation for it and dealing with it as if it is a reality. Nowhere is this so striking as among the religious groups that occasionally emerge and announce they will sell all their worldly belongings, bid farewells to their friends and relatives, and await the end of the world a week from Thursday.

This swirling river of human reverie has to be accounted for and explained. Is it possible that if the brain doesn't imagine actions, people won't act? Or as scripture puts it, "Where there is no vision, the people perish."

We are not addressing philosophical questions such as what is real and what is represented in the brain. They have occupied the likes of Descartes, Locke, and Berkeley, as well as contemporary scholars. They remain largely unanswered and the non–game show carries on. Genuine contentiousness continues.

But we are concerned with how the brain works *without* people working at making it work. We wonder how brain work is experienced directly without a third eye looking on. From this perspective, what is imagined can be as real as biting unexpectedly into a red-hot chili pepper. For example, hell.

IMPLICATIONS?

Historically, the distinction between the experienced and tangible and the imagined and believed has not always been clear. Nor is it always clear today. For example, when the idea of humans flying was first proposed, it was thought to be an act of imagination. But then it happened. Or, as with witches, there were those who believed they were real. There were well-documented lawful witch hunts in Loudon, France, and Salem, Massachusetts.[22] Condemning and sentencing countless accused witches in countless courts and then burning and killing them was relentlessly real. And, to no one's surprise, this week and last there have been new sightings of flying saucers, endless new conspiracy theories, and at least three dozen new bogus medical cures that thousands of people will believe. Once more, somewhere there will emerge imaginings among some particular ethnic or national or religious group that they are members of a superior race or faith. Superior—

full stop! An absolute and permanent hierarchical success. Nazis. The folks at Jonestown. The woman who detonated herself and killed Indira Gandhi.

A technical term describes imaginings that assume the quality of being real—*attribution*.[23] It means attributing or assigning meaning or a causal explanation to tangible events and experiences as well as imaginings.

And it's commonplace. We all do it. For example, people attribute certain features to heaven and personalities to deities. They project themselves into an afterlife, oftentimes with very specific details such as the hour at which breakfast is served and what's on the menu. Saint Peter jokes usually depict a place much like an exclusive upscale American or Saudi suburb. There is a gate and a gatekeeper. One must qualify to enter and not everyone does. Inside there are important and deserving people. And on the top of the hill is a big mansion that houses the boss of all bosses.

Or in daily life someone prays to his god for the health of a seriously ill loved one. Over time her health improves and it's perfectly easy, even reasonable if not axiomatic, for believers to conclude that the prayers were somehow instrumental in mediating the outcome. Surely it happened because the attributed deity with an attributed set of values somehow arranged it. Of course.

But say this ends in death. It's perfectly easy amid the turmoil of grief to believe that for some reason the deity decided it was the loved one's time to change her current address to one in heaven. In both cases the hierarchical primate concludes that someone must be in charge, with quite the same degree of confidence as his assurance that things don't fall up. There has to be a hierarchy, and within it a functioning agency. As Daniel Dennett has remarked, "We have a built-in, very potent trigger tendency to find agency in things that are not agents, like snow falling off the roof."[24]

Different parts of the brain, as we come to later, enact these various scenarios. Fascinating as they are, they resolve themselves this way: for whatever the reasons, the brain has evolved as it has, and whatever else it does, it imagines and believes and acts on its beliefs. Beliefs determine a person's actions and what a person expects will be the consequences. This is clear enough among committed believers, but then how many *on-the-fencers* have joined a religion just in case, impending death being the catalytic converter? And religions—*all of them*—are clusters of beliefs with very special characteristics. They may vary like flowers, but they are all beliefs.

As we have already stressed, the most prominent of them is belief in an imagined higher authority or an absolute and personified moral principal that guides behavior. Every group member is also a politician willing and able to accept a negotiated position in the larger scheme of potent order. It is apparently very difficult for people to accept that they are alone in the universe and responsible entirely for what happens to them. They are also unwilling to accept that their final leaders are also without someone still higher in the universe. The last line of the presidential oath of office is usually "So help me God," which means, even you, new big shot, have to listen to someone.[25]

Some people try to operate with their private map, and they relish the challenge and sport of it. But we will come to discuss how we evolved as a social species. Choosing an antisocial or marginal position requires special energy and gifts and very often some alloy of recklessness and courage. Or so the record of the world suggests.

BELIEF MAN AND THE RELIGIOUS RESUME

But the story of religion doesn't end here.

We can, for clarification, isolate four fundamental pillars of religious operation: sacred texts, deities, dogma, and rules about behavior.

All religions tell a story. They are found in the oral history of the Hopi, the Torah and Siddur, the New and Old Testaments, the many texts of the Gnostics, the compiled teaching of Buddha, the writings of the Sufis (Islam), the scriptures of Tao and Taoism, the reflections of Bahais-Bahauddah, the Book of the Latter-Day Saints, and hundreds more. Equally interesting are the derivative fictionalized stories about these celebrated stories, such as *The Da Vinci Code*, of which 40 million copies have been sold, presumably because it evokes both the powerful drama of ancient religion as well as its setting in modern daily life. Sacred texts, deities, dogma, and fierce or broken rules about behavior are religion's story line.

For each religion, there is usually a single, broadly accepted official version of sacred text, although at times more than one version is accepted. For example, not all versions of the Bible are acceptable to Episcopalians. And there is usually a divide between those who are privileged to interpret the texts and those who have to abide by the interpretations. Texts sometimes contain information about sacred meeting times—the Sabbath—or sacred places—temples. There will usually be rules for certain events such as the Hagadah, which functions as a program for Passover. There may be rules of behavior that apply both in and out of sacred places and independently of sacred times, such as the Ten Commandments. They let you go anywhere and follow you there, like a passport. For those who worship or depend on animals or the natural elements, the promise could be personal or group safety, or communications with deceased ancestors, or good crops during the coming year. And if the crops failed or did poorly, at least the individual farmer enjoyed the ability to claim the situation was not a consequence of ineptitude or haste but of larger forces within his own, perhaps prickly, system.

The list of the major holy writings intrinsic to religion is long and rather amazing. The authors of sacred texts are nothing if not imaginative polymaths.

With the exception of Buddhism, there is almost always one or more imagined deity, one in Christianity and Islam but many in Shinto and Hinduism. These are powerful beings, special forces that preside over events on earth and elsewhere. Each of these makes particular claims and has different expectations and rules about behavior of those on earth. Each of these has a different message of hope and promise. Each invites personal commitment. The greater the degree of true commitment, the greater the likelihood of a special personal relationship with the deity or force, even if it's through an intermediary, such as one's saint in Catholicism. One doesn't pray to the pope or the imam or the rabbi, but to one's deity. Judaism forbids worship of any concrete person or thing other than the deity, who cannot be depicted. Imprecision or generality are evidently regarded as advantages or virtues. Especially for nomads, fixed icons are a costly luxury.

Religion's stories are attractive. Their messages are powerful and often nothing less than mountainously mighty stories, events, admonitions, and promises. They invite belief and, as noted, at times spellbinding involvement. And they serve as an ominous warning to those who would risk not believing or dare to interrupt a sermon in the face of danger.

There is yet another side to belief and dogma. We have stressed that people can't avoid believing things they imagine. Such believing is associated with a frequently ignored bias of the brain to perpetuate one's beliefs. When it is applied to religion, this bias serves to reject ideas and evidence that conflict with what is believed. Most Christians reject the possibility of no afterlife. For Jehovah's Witnesses, it's the possibility of deities other than Jehovah that is rejected. For creationists, the list of their rejections include geological evidence of the age of the earth, tangible evidence of very ancient human ancestors that discounts the creation story found in Genesis, rock art paintings depicting religious events dating back forty thousand or more years, changes in species as tracked by evolutionary theory, and—for imaginative souls—the reality of death, which is believed to be but a step onto an escalator headed for the infinity above.

We haven't emphasized unduly—it's unnecessary—that committed brains often develop novel, if not ingenious, explanations when predicted events don't occur on schedule, such as the ending of the world seven weeks from today. Somehow it turns out that the dates were wrong and new and improved pronouncements speedily offer newly embraced beliefs about a new final hour. Occasionally this ploy goes tragically wrong, as in the mass suicide at Jonestown. But overall, it is instructive that the dire recipes of systems of belief lead to so few rather than so many outright fatalities. Of course, here we must note the loomingly large exception of religious wars and the special and often successful urgencies of suicide bombers for whom belief is a cluster bomb.

There are two clear features of the pillars of religion we have described that carry implications for how religions operate. The first is that beliefs, dogma, texts, and depictions of deities need to be crafted so that they don't tie too closely and perhaps inconveniently to reality. What religions abhor are well-controlled scientific studies that, for example, show that serious and repeated prayer for ill individuals does not improve their health.[26] The second is that once the brain buys into dogma and the reality of a higher authority, it is unusually vulnerable to deception, particularly regarding text interpretations, spiritual sanctions, and rules about behavior. It has become self-programmed. Were it a Microsoft program, it would require frequent patches to remove its often-subtle malfunctions. People are slow to leave religions and their ways of explaining life simply because it's difficult to migrate to a different religious neighborhood, even if the main visible difference is that one worship focuses on Friday and another on Saturday and another on Sunday.

There are of course distinct differences in the ways in which dogma, text, and rules are believed and what they are supposed to accomplish. There will be ongoing issues, such as ideal ways of praying; for example, your forehead has to touch the stone floor when you bow in the mosque. Or your prayer is for nothing if you are focusing on the lady's hat (or the lady) in the pew ahead. Even what one should and shouldn't eat and how one relates to members of other religions are entered on the religious resume. Muslims view the Qur'an as the direct word of their God. They aspire to pray five times a day to assure that their behavior is consistent with scripture, just as they aspire to make at least one holy pilgrimage to Mecca during their lifetime. It's all clearly dictated in the book. On the other hand, some Christians view the New Testament as an interpretation of God's message and historical events, not his exact words. But they have invented special and remarkable behaviors, such as kneeling before entering a pew to demonstrate evidence of obedient religiosity. Other Christians view the Old Testament as the truth and the New Testament as of secondary importance.

To this can be added the stories of influential interpreters of text (Benedict of Nursia, Maimonides, Saint Francis) and critical moments at which change took place, such as the adoption of the Trinity by Catholicism. And there are different end points and proscriptions about how to achieve them. Compare Buddhism with American evangelism. Buddhists discover their true nature by a personal, *private*, spiritual, mental journey— primarily meditation—about which others may be unaware. In evangelistic circles, the journey is as much public as it is personal and private. Enact good deeds, contribute financially, behave as suggested in sacred texts, obtain new church members, work on church functions—these are the assignments. Good works are the visible requirement for redemption. The eye of the needle is well-lit by a brilliant spotlight.

And now, Paradise!

Lost? Misplaced? Where?

Religions are typically characterized in terms of their dogma, physical structures, and sacred places (e.g., Mecca, the Vatican, Jerusalem), their practices (e.g., don't eat pork, fast on Saturday, confess sins on Sunday), and their stated intentions (e.g., bring the unwashed and unsaved to god, who is also to be celebrated). Some of the beneficial effects of these activities, such as the care of their own (as with the rescue activity by monks of stricken Burma during the late spring of 2008, despite the foul government there) have been mentioned, with more to come.

OTHER PARTS OF THE STORY

For the moment we will switch our focus elsewhere. Many religious activities are not quite as innocent or pure or as beneficial as they might first seem or are claimed to be. Scientology is regarded in some jurisdictions as a religion but as a financial racket in others. Often nonreligious groups, such as corporations and governments, use religions for reasons of state or they manipulate religious dogma and sacred rules to their electoral advantage. The distinction between church and state is often difficult to recognize. In fact, many religions and states aggressively refuse to acknowledge a distinction even exists.

Examples from politics and finance can illustrate.

Nations such as China and the United States argue heatedly over religious intolerance. But at the same time the Vatican and China matter-of-factly explore the possibility of normalizing relations that were angrily severed in 1951. A branch of the Presbyterian Church publishes a 9/11 book that posits a CIA conspiracy as its cause.[27] The US House of Representatives injects prayer into the 2007 Defense Bill.[28] Muslim clerics criticize anti-Muslim cartoons by European newspapers and physically threaten Pope Benedict XVI for his statements about medieval Islam. American evangelists criticize other nations, their leaders, and their religions. With a vigorously imaginative notion of causality, the Reverend Jerry Falwell described the bombing of September 11 as "divine retribution" for American homosexuality and abortion. Students at a number of American colleges refuse to pledge allegiance to the American flag and to utter the word "God." On the other side of town, conservative Christians call for consumers to boycott retail outlets that refuse to use the term "Merry Christmas." Conflicts between those favoring stem-cell research and those who, because of their faith, oppose it despite its potential for reducing disease continue to shape the ever-changing landscape of bioethics.

There are also, and always, finances. The most fundamental is a request for members to contribute financially for ongoing costs. The hierarchy must, after all, be supported, even though a church may have vast holdings of rural and urban real estate as in the case of the Roman Catholic Church in Europe. But there are also many religion-sponsored public moments of commerce with raffles, lotteries, books for sale, and tax-deductible gift opportunities. Americans contribute more funds to religions than to any other sector of society. Religion can readily seem to resemble big business, even if the profit motive is not its primary one but cash flow is. A rather detailed description of the organizational and financial basis of the evangelical appearances of Billy Graham suggests the intricate practicality necessary to support the most lustrous of religious events.[29]

The laws and strategies of economics apply just as much to organizations of the reversed collar as the one that is buttoned down.

Later we will discuss proselytizing that is in a fairly direct way the religious equivalent of business expansion and seeking market share. Just as there are business scandals about undue use of corporate resources or stock option manipulation, among religions there are ample examples of church fraud and leaders' savagely luxurious lives funded unknowingly by church members.[30] And even those outside religion often join the party. For example, the University of California at Los Angeles has confected a $42,000-a-head, twenty-one-day, no-host bar tour by chartered jet to seven major historical religious sites.[31]

These points should not be taken to suggest that religions don't have positive features. Our central claim is that they do. For those in their sway, religions do have positive effects right in the neurophysiological core of consciousness. Endless testimonials attest to this fact. So too do the raw numbers of followers, which number in the billions. Most religions make allowances for individual differences in piety and performance. Remarkably, Catholic priests hearing confession will often forgive and bless the parishioner who repeats the same sin each week. In Islam, if one is poor or there are extenuating family circumstances, there is no expectation of pilgrimage to Mecca. Ardent evangelists show a respectable degree of tolerance by trying to encourage their on-the-fence members to take a leap of faith but will share a beer with them even if they don't. Members of the same church help and pray for each other. Churches open their doors to the homeless and provide food and shelter. During disasters such as Hurricane Katrina religious groups voluntarily contributed space, time, money, and hands-on assistance to the afflicted far more effectively than the governments paid to do so. A number of conservative American religious groups have entered the current crusade over climate change.[32] And how many are aware that the Catholic Church is the largest single supplier of healthcare and education on the planet, a principal glue of civil society in Africa, and the strongest opponent to the caste system in India?[33]

We have tried in this chapter to capture the complexity and consistency of spiritual and religious impulses and to stress how vital it is to approach them from an informative perspective.

We are dealing with what exists, not what we may want more or less of. Bitterness about the laws of gravity or hydraulics has never been helpful to those who design airplanes, sculpt exquisite irrigation systems for growing rice in Cambodia, or move tons and tons of earth and stone to build robust dams along the Colorado River. Anyone wishing to change a system or maintain and strengthen it had best understand it first.

Chapter 3
ADVENTURES OF THE SOUL

In this chapter, we dial down even further. We are particularly interested in the degree to which religious beliefs and rules about behavior permeate the lives of believers, day in and day out. Whatever the battles that surround religion and truth, everyday lives are the front lines. This is the area of personal biography.

As with full-time jobs, there is a dailiness to religion. For committed believers, and especially those who embrace their religion with an ideological totalism, their religious beliefs and rules are as much a part of their lives as their pancreas. And far more clearly, when asked who they are, not many people (if even one) will describe themselves as the possessor of a pancreas. But they are very likely to use their religion as a defining or at least primary characteristic. This becomes more salient and significant where there is more than one religious group in an environment, such as the city of Jerusalem. Then the affiliation becomes more important and even mandatory. It can also become potentially fatal, as in the Sunni-Shia distinction of contemporary times or the Catholic-Protestant divide in Northern Ireland or the centuries of religious wars in Europe.

It is ubiquitous, unremarked upon, and amazing but also clarifying to remember that countless people daily wear on their physical bodies amulets or signals of their spiritual commitments. The power of the world of spirits becomes symbolized by a small object of stone or precious metal or artful fabrication.[1]

Even when the spiritual and moral implication of a religious identity is to an individual meaningless at best and irritating or infuriating at worst, the identification remains. It is adhesive. For example, author Tiger was born a Jew in Montreal, a city he defines as one in which everyone is a member of a minority group, including the French-Canadian majority. Religious identification there was not solely spiritual or pious. This was especially so during the Second World War (the period of Tiger's childhood) when there were riots against the conscription of French-speakers to fight in the war on the side of the English Crown, which had defeated Quebec in battle hundreds of years before.[2] It was also seen by an influential number of Quebecois as an effort to protect the heretical Jews who were not only stereotypically seen as economic exploiters but also, and more gravely, the tribe responsible for killing Christ. So the matter of what one was was not theoretical. It was almost visceral. Bodily survival or welfare seemed a genuine issue and when the young lad walked down a street controlled by enemy boys, such as the dreaded Jeanne Mance Gang, hairs on the neck literally stood on end in fear.

However, the theology that supported Jewish identification was never to equal the trepidation about physical safety. Young Tiger was subject to the customary ritual of bar mitzvah at age thirteen. He experienced appropriate and desirable enthusiasm for the entire solemn folderol. After all, it was surely very consequential. It marked a change from childhood to responsible and serious maturity. A different relationship with God was supposed to follow. No longer would Father Tiger bear responsibility for his son's flaws and generous acts. Now God would judge them directly and reward Tiger Junior's efforts to obey religious law, perform self-discipline, and please others. To buttress this epochal accountancy and gratify God, he recited ritual prayers not only for the mandatory and central bar mitzvah matters but also for such arcane observances as wearing a new shirt for the first time or enjoying the first fruit of the season. These were, after all, in the manual of instructions for how a grown-up lived and worshiped. And the ceremonial activity at the family synagogue proceeded with dignity and evident success.[3]

Earnest though were his intentions and however reasonably thorough was his loyalty to the new adult tasks, the process failed to take. After a respectable length of time in which young Tiger sought with interested eagerness a sign of enhanced attention from divine force, the lad reluctantly and dolorously had to conclude that there was no credible evidence of greater godly interest in him than before. This was despite his willingness to embrace enhanced responsibility. There emerged no acceptable evidence of a personal and definable kind that would certify and thus cement the newly adult relationship with the higher power. Indeed, the loss of the expectation and desire for faith had for Tiger Junior little of the drama and literary color of anguished loss-of-belief experiences as recorded by countless predecessors. Of course, he continued lively interaction in the Jewish community within which he was embedded. However, the full religious element diminished in significance.

We have sketched some of the permutations of how religion permeates daily life. We discuss them further in later chapters. Some are common knowledge. Key beliefs, texts, founders, and behavioral rules, such as those governing food and dress requirements and typical rituals, are well known even though they are frequently based on very theoretical, abstract, and seemingly arbitrary assumptions. People may know them better than the names of their representatives or senators, the capital of Alaska, or the most recent pitcher of a no-hitter on their local baseball team.

This is hardly surprising. The world of faith is imposing and complex. Religions differ, sometimes seemingly rather minimally, such as between Methodists and Lutherans, who are to outsiders at least more similar than not, even though adherents of each faith will not accept such a casual diagnosis. Catholics and Buddhists and Jehovah's Witnesses, on the other hand, are more different than similar. The variation among the world's thousands of religions is dazzling and absorbing.

Depending on which religion is under the microscope, different facts and profiles appear. We have not tried to review these differences in detail. This has been done many times before and very well indeed. Instead, here we offer three vignettes in an illustrative but not definitive effort to capture the impact of religious influence on three people. We describe a typical week in the life of an evangelistic American male. We also examine the observations and prayers recorded in the journal of a young Catholic female in a covered wagon crossing the United States in the 1870s. And we describe the fate of a young child who received a thorough Catholic indoctrination in Nazi Germany, only to later learn that he was born a Jew. The first case involves a person one of us knows who experienced danger without leaving home. The second concerns a member of author McGuire's family and her story is of a perilous covered-wagon trip across the United States. The third case is a composite of individual stories for reasons clarified at the beginning of his tale.

Before turning to these real-life examples, let's engage in a made-up experiment in conjectural and improbable thought.

Let us assume that for some reason presumably connected to the nature of aliens, a personlike individual has arrived from another planet with a self-imposed mission. He, she, or it seeks to become a believer. The creature wants to adopt a religion and to be adopted by it. He is a thorough and fair soul, and in order to decide which religion to embrace, he earnestly interviews people across North America.

We share his experience.

He visits the recently constructed and opened Creationist Museum in Petersburg, Kentucky.[4] He notes that its major theme is to challenge evolution. Thus he suddenly finds himself pondering a controversy that was strange news to him.[5] He has to stop by a doctor's office for a moment to obtain some medication for an unexpected allergy, and in the waiting room he reads that most doctors view religion as beneficial for their patient's health and longevity.[6] During a tour of a library, which he has been informed is where native knowledge is stored, he reads an article by a scholar of religion that compares the current chaos in the world to the Reformation and its central role in keeping Europe's political house divided between 1500 and 1700.[7]

The story becomes still more complex. At an evening lecture series, one speaker proclaims religions exist because they provide existential security by combating the fear of the unknown, while another speaker claims that Jesus was a prophet, not the son of God, and it was his disciple Paul who was responsible for fashioning a religion around the figure of Jesus.[8] Outside the lecture hall there is a rally by Jews-for-Jesus. Further down the street, a group of Baptists accompanied by a jazz band are engaged in an open-air, everyone's-invited revival.

At a prestigious academic center, a confident philosopher baffles our alien by informing him that religions are an attempt to identify the rational order beneath the deceptive surface of daily life. A geneticist asserts that there is a god gene that predisposes people to seek out and worship a higher authority. A historian maintains that Luther, Calvin, and Swedenborg suffered from an obsession to identify the footpath to an ideal life. Still another historian asserts that the only true religion is Judaism, while a historian in a nearby office argues that anyone who believes the Old Testament ought to have his head examined. A sociologist explains why religions are the cause of wars. A physiologist announces that serious religious meditation improves one's capacity for attention. And a neuroscientist shows the visitor pictures of the brains of nuns living in Montreal, photos pinpointing specific functional brain centers that increase their activity when people pray or allow "God to speak through them."

Our visitor mobilizes his Infinity Express Gold credit card to visit the United States Air Force Academy. He assumes he will see there the essence of technological and secular human skill. After all, students at the academy learn how to control aircraft speeding at 1,600 mph while they aim and fire missiles that can find a six-foot target miles away. Nevertheless, despite the genuine technodazzle, he encounters equally genuine open primordial conflict among chaplains of different faiths.[9] He hears complaints of discrimination by chaplains of minority faiths. Over dinner, he talks with a member of People for the Ethical Treatment of Animals (PETA) who happens to be a pilot of supersonic jets. The member assures him that PETA is a religion and that its members advocate sacrificing humans, not animals, for medical research.[10] On someone's desk he sees a letter from one air force officer to another with the sign-off, "Yours in Christ," a salutation wholly impermissible by the formal rules of air force activity. After dinner, he joins a conversation in which the participants are questioning whether religion is part of human nature.[11]

His wanderings continue. Now he is in Salt Lake City,

where he encounters Mormon disciples of Joseph Smith Jr. They warmly attest to the truth of revelations described implacably and with a mystifying and anesthetic lack of drama in the Book of Mormon. He is shown confirmation of the fact of Joseph's revelation when an angel visited him in 1823.[12] He is told about a settlement of Mormons who, openly and with full theological confidence founded in the Book of Mormon, employ the practice of polygamy, long defined as illegal in the United States. He is told their practices are protected, or should be, because of an American commitment to freedom of religious observance and that researchers have found brain differences between believers and nonbelievers. Nonetheless, polygamy represents a substantial challenge to religious tolerance in the country.

Throughout his travels, he notes an astonishing array of religious goods and services, such as teaching materials, prayer books, trinkets, and devotional objects. Of particular interest is a book kit intended for soldiers to protect their sexual purity.[13] These flow as if from religion's assembly line and find their way nearly everywhere, from talismans on the skin to candles shaped as pious figures. At a supply store for religious materials, he is informed by a salesman that "George W. Bush was not elected president of the United States by its citizens but placed in the presidency by an act of God."[14]

In each town, he observes many well-intentioned and often inspirational and humble members of the cloth. These persons are totally convinced of the presence of a higher authority. But they believe, too, that it is their self-evident mission to make their special knowledge known to others. They regard it as their calling and their duty to convince others of the implacable correctness of what they preach. He learns that some of these people even travel halfway around the planet to pursue their mission, often in situations of considerable discomfort and even danger, both political and physical.

He is struck when he observes that many of those who are going to or departing from religious services seem unusually happy and relaxed. Something substantial seems to happen in that building or on the meadow. A visit to a primate research center reveals that many scientists believe humans and some nonhuman primates share biological systems that are the basis of morality and, possibly, very rudimentary religion.

With his travels almost finished, he visits as many different congregations as possible. But he has to conclude that most embrace different systems of thought that are housed in separate organizations that hardly communicate with each other. Seemingly no one cares, though there are evidently sporadic episodes of hostility that flare up for an array of reasons and then subside.

He notes that sermons and teaching often focus on moral views of right and wrong rather than the obvious reality he perceives of human suffering and discrimination. "Who is religion really for?" he wonders. A final encounter with a minister introduces him to the plausible and congenial view, that, despite their differences, all religions are really only separate paths to the same goal. However, he then encounters the bewildering caveat from an official of one church, that only female virgins are completely assured a place in heaven.

Exhausted and confused, our visitor wisely decides against extending his search to Asia, the Middle East, Africa, and South America, though many enthusiasts have boasted of the historic and still burgeoning success of ardent missionaries. He is convinced, or at least convinces himself, there is little prospect that such a trip could provide the decisively novel information he needs to make a decision.

What might this visitor now do to become a believer and fulfill his mission? How would he decide which religion, if any, to adopt? Which of his experiences, if any, would affect his decision?

He could of course return to his place of origin after concluding that there is no way to make an informed decision—but for a thought experiment this solution is hardly fair.

He could pause, review all his notes, update them with further research and interviews and then decide. But this is unrealistic. In truth, perhaps one person in a thousand engages in a serious in-depth comparison of religions before committing to one.

Or, in desperation, he could join a religion that outwardly appears attractive or is the only convenient and sociable option. This is, in part, the story of Robert, which we will tell shortly. Or he could have arrived at a much younger age, lived with a family of believers, been gradually assimilated into the family's religion and thereby become a believer through a leisurely osmotic process. This is, in part, the story of Elva, we will later tell. And then we will hear about the more complicated plight of uprooted Jewish-born Catholics confounded by the lack of clarity of their religious identity. A lack of clarity, it should be remarked, about a matter steeped in lack of clarity and in oceanic imprecision. Perhaps they somehow expect that everyone should have a religion that is crystallized and fully formed. But where does such an idea come from?

ROBERT IS AN EVANGELIST, AMERICAN MALE

The amount of time people spend in activities either expected or recommended by their religion varies.

At one extreme are those who profess that they have a strong religious affiliation. But they rarely attend religious services or engage in religious rituals or participate in religion-related activities, such as community service. They pay only minimal heed to the advice and the ruminations of religious authorities and spokespersons.

At the other extreme are those whose lives unfold in a religious context and with obvious religious meaning. They may live anywhere. But many live in rural or exurban areas in America where the church is an area's social center and serves myriad functions other than providing sermons and the opportunity for repenting on Sunday morning. While suburban and city folk are spending time at movies, attending sporting events, shopping, walking in parks, or engrossed in esoteric hobbies, rural folk often live where there is no movie theater within easy reach, where sporting events are limited to an occasional nearby high school football game, and where there are no shopping malls within less than a half-day's drive.

Robert lives in rural America, in a town of approximately 1,600 people. There are two gas stations, one small market, one eatery sporadically and inconsistently open or not, one car repair garage of unreliable reputation, and one bar. He is now twenty-nine years old and a carpenter by trade. He occasionally drinks wine but not hard liquor. He doesn't smoke or chew tobacco or sniff snuff. He is financially solvent with an agreeable balance of needs and resources.

Robert's story is not wholly unfamiliar. His parents were highly religious and active members of their local church. While he was young, he attended church services and helped with church restoration projects. None of these activities inspired him, or they seemed little more than a waste of his time.

School was moderately interesting until he was old enough to own an automobile. At age sixteen he dropped out of high school and took a job in construction. Events soon took a turn for the worse. By age twenty, Robert had been convicted for the robbery of a local gas station—fifteen months in prison was his reward. Two citations for drunk driving over the next year only added to his misery and obvious lack of direction. At age twenty-four, he was out of work, penniless, unable to find a job, in debt, and wandering. He returned home, but his parents refused to let him live with them.

Then one day at the local garage he talked with a friend from high school who told him of his conversion to Christianity and how it had changed his life. He encouraged Robert to visit his church. Robert agreed. What follows is a week in Robert's life, a year and a half after he first visited his friend's church.

Sunday: 30 minutes in travel from home to church / 60 minutes in attendance at the first Sunday morning service / 90 minutes help with Sunday school / 60 minutes in attendance at the second Sunday service / 15 minutes informal discussion with church members / 30 minutes in travel from church to home / 60 minutes in private Bible study in the evening / 20 minutes in private evening prayer. Total = 365 minutes.

Monday: 10 minutes in morning private prayer / 20 minutes in evening private prayer. Total = 30 minutes.

Tuesday: 10 minutes in morning private prayer / 20 minutes in travel from home to Bible class / 70 minutes in attendance in Bible class / 20 minutes in travel from Bible class to work / 20 minutes in private evening prayer. Total = 140 minutes.

Wednesday: 10 minutes in morning private prayer / 45 minutes taking an invalid church member to and from her home to the market and assisting her in shopping / 20 minutes in evening private prayer. Total = 75 minutes.

Thursday: 10 minutes in morning private prayer / 30 minutes in travel from home to church / 60 minutes in a discussion group of male church members / 20 minutes driving from church to work / 20 minutes in evening private prayer. Total = 140 minutes.

Friday: 10 minutes in morning private prayer / 30 minutes travel to a social dinner for church singles / 120 minutes at the Singles Dinner and singing following dinner / 30 minutes in travel from church to home / 20 minutes in evening private prayer. Total = 210 minutes.

Saturday: 10 minutes in morning private prayer / 30 minutes in travel from home to church / 150 minutes doing carpentry in a church renovation project / 30 minutes in travel from church to home / 20 minutes in evening private prayer. Total = 240 minutes.

Total minutes spent in church-related activities during the week = 1,200 minutes, or 20 hours.

No doubt twenty hours per week devoted to religion will seem extreme to many readers. But following his conversion it was by no means excessive for Robert. He had grown up in a religious family—all were and remain active members of evangelistic churches.

Belief is clear. The existence of God is a fact. The Resurrection of Jesus is a fact. The writings of Matthew, Mark, Luke, and John are Gospel. "God deserves everything we can give him. . . . He is our inspiration," were the words of Robert's mother as she began the half hour of Bible reading for her three children each day before dinner. The entire family had worked on church projects. They had hosted and attended church social dinners and other events. They had decorated the exterior of the house for Christmas well before Thanksgiving. The minister frequently visited their home. Robert's father gave 12 percent of his earnings to the church. Robert's mother gave half of her $50,000 maternal inheritance to the church. At age twenty-six, and at last a confirmed believer, all this suddenly made sense to Robert. In addition, he liked the people at his church. "They make me feel important, an essential member of the community, and they soften the rough edges of long days at work."

Conversions such as Robert's are not unusual. They have a long and influential history dating back at least to Paul and Augustine.

Robert also found out something interesting about himself. A year earlier he had contracted to build a cabin in a remote mountain site. For the entire project he was alone. Only on Sunday would he leave the work site and drive fifty miles on dirt roads to the nearest church. But there he found little satisfaction. Sermons were all that was offered: no social events, no Bible reading classes, and no community activities. After seven weeks, the job was done and he vowed to never again separate himself from his church and its people. "I'm more social than I realized," he told his minister. "Only in the presence of the people of God and when I give to others do I feel right." In his giving to others perhaps Robert was experiencing what neuroscientists have reported recently: acts of generosity light up the same reward centers in the brain as food and sex do. And prison officials and scientists alike know well that solitary confinement is a drastic and debilitating punishment.

Of course, not all of Robert's religious day was strictly religious. People at church discuss politics, jobs, and prices, exchange cooking strategies, and update each other with local news. But also, often, the church and some of its many activities are the topics of discussions away from church. For example, a missionary and his family return from Africa and tell the story of their physical dangers, hardships, and triumphant conversions. Their story grips the imagination of the church members as well as the town's few nonbelievers. It takes on a life of its own in the community.

Pause here a moment and ask: how is it that a person closely similar to Robert—imagine he had a twin brother who as a child went to live with relatives in Baltimore—can spend literally zero minutes a week in religious activity while Robert spends more than two full working days? Psychologists claim that the propensity for spirituality is highly similar among twins. But this explanation would not seem to apply to Robert and his imaginary twin. Is it the social environment in which people live that overrides this propensity? Probably, or at least often. Especially in circumstances such as Robert's, where the local church is also the town's social center. Urban areas offer alternatives. Yet it is in cities that Protestant evangelicals are more effective (for example, now in South America and at one time in Korea). Or is the type of religion the deciding factor? Probably not. In many rural areas of Tibet, South America, and Africa different religions prevail. But the local place of worship serves similar functions to Robert's church and the majority of adults in the area attend. Relatively few choose self-exile.

What then might Robert's new life have offered him that he may not have recognized? We spell out an answer in forth-coming chapters. But a preview or a literal heads-up can be provided here. Positive socialization—that is socialization characteristic of religion-related events, enacting rituals, and incorporating and committing to religious beliefs—predictably *brainsoothe*.

Religions excel in *brainsoothing*.

Epilogue. Robert is now twenty-nine years old. He is mar-ried. He and his wife have a child. He is a building supervisor for a company in a nearby county. And he spends about fifteen hours a week in church-related activities.

ELVA TAKES A JOURNEY ACROSS THE UNITED STATES IN THE 1870s

Here is the story of another journey. It is one of the epic, often heroic and legendary ones that Americans endured as they traveled across the continent to the West Coast or as close to it as they could get. The principal description of it derives from discussions and a journal kept by a female relative of author McGuire. It highlights the importance of religious belief to an individual and a group of people under drastic and enormous pressures of uncertainty and scarcity.

The groups traveling west during the last half of the nineteenth century varied enormously, from idealists seeking a new utopia to gold rushers to socialists to religious adherents such as the Mormons. Our case in point here was composed of Catholics, Unitarians, socialists, and labor organizers with whom Elva traveled west from the coalfields of Pennsylvania to Jerome, Arizona, in 1876.

She was born in Vermont. By six years of age, she was an orphan after the sudden death of both of her atheist parents. She was adopted by an aunt in Pennsylvania. Aunt Millie was a devout Catholic. Elva soon found herself baptized and an active member of the local Catholic church. Time passed. Her school performance was exceptional. Teachers spoke about her attending college. Local males proposed. At age seventeen she was accepted in a nearby college and was preparing to attend when Millie announced that she, her husband, Felix, other family members, and some friends were "moving to California for a new life." Millie was aging. Elva feared that without her assistance Millie might not live to see California. Elva joined the group.

The trip took nearly seven months. From the beginning there were disputes over religious beliefs and rituals—for example, who said grace before supper or if it was said at all—and leadership. Making decisions about routes, how far to travel over a day, and who was responsible for which chores was never easy or automatic. Yet nothing in Elva's experience was atypical for the time. It was a period when other humans were the only resource during travels across America. There was hardly any government, no automobile tow services, very few hotels, only food of questionable quality, and no cell phones or Internet. The tool kit contained a compass, hard work, compassion, motivation, inspiration, and hope.

In Elva's interpretation the group got as far as it did—Jerome, Arizona—because, "God looked over us all the way." Slightly edited selections from her journal and discussions with author McGuire follow:

> *Kentucky.* "The rain has not stopped for three days. All is wet and no one can set a fire. . . . No use trying to move the wagons. . . . It's a world of mud. . . . Millie has a cough and it sounds serious. . . . A horse died last night. . . . I will pray tonight for a change in the weather and Millie's health."

> *Two days later.* "The rain has stopped. . . . We are moving now. It is still muddy. . . . Millie is coughing less. . . . God answers prayers."

> *Arkansas.* "Everyone is sick, coughing, and spitting. . . . I was the only one strong enough to cook dinner last evening. . . . The men are still arguing over who should lead. . . . Mr. Ervin (a labor organizer) wants a vote. Felix says he should lead because he is best qualified. . . . I feared of a fight among the men, but it didn't happen. . . . We were very anxious when Mr. Ervin said he had seen Indians. . . . Maybe prayers will help—God will know who should lead and give us health again, and I will ask him to keep us safe."

> *Next day.* Felix assumes leadership of the group.

> *Five days later.* "Most everyone is now healthy, thanks to God."

> *In what is now the state of Oklahoma.* "It's water we need. We need it desperately. It has been three days since any

water. . . . The heat is intolerable. . . . The horses are tired and won't move. . . . Two axles are broken and need to be replaced. . . . Everyone stays in the shade. . . . Felix is guarding what water is left—he says perhaps enough for a half-day. . . . We have come a long way. . . . Will God desert us now? . . . I will tell him of our plight."

Next day. A thunderstorm drenches the group. Humans and animals forget the wetness as they stand in the rain and fill their cups.

Four days later. With water aplenty and broken axels replaced, the group heads for Texas. "God is kind and forgiving."

Six days later. They stop to bury the remains of another group whose members died from starvation and lack of water.

Texas and what is now the state of New Mexico. "Millie's cough has returned worse than ever. . . . She will only take broth. . . . Felix has stopped travel several times today to give her a rest. . . . I am afraid the trip is killing her. . . . Everyone is tense. . . . I will pray."

Ten days later. Millie dies and is buried. "Dear God, take Millie into your home. I accept that you decided that her time had come to join you. She is a wonderful woman and was a wonderful mother to me. Tell her that when my time comes I will join her. Oh God, forgive me for this selfish message, but if not for Millie I would be wandering aimlessly without knowledge of you."

The town of Jerome in what is now the state of Arizona. "We have dissolved. Mr. Ervin and three of his group took their belongings and left. . . . There were harsh words between them and Felix. . . . Felix has decided that the rest of us should stay here through the summer and then start for California in the fall. The Unitarians are holding their own prayer meetings but so are we."

That evening. "Dear God. Thank you. You have been with us from the beginning of this trip months ago. We are here because you have blessed us and seen to our survival. We love you."

Epilogue. Elva and Felix remained in Jerome. Elva took a job as a cook in a mining camp. Felix became ill and died. Three years later, Elva moved to California, married, had one child, lived to age ninety-six, and remained a devout Catholic throughout her life.

JASON: A CHILD OF HISTORY AND SURVIVAL

Jason is a composite, created from parts of the lives of several individuals in order to minimize the possibility of identification.

During the 1960s one of us (author McGuire) was working in a hospital in an East Coast city and had the opportunity to interview several individuals with similar histories. The story of Jason is an alloy of their stories.

It is the early 1930s in Germany soon after Hitler had assumed leadership of the country. For many Jewish families it was soon apparent that their civil rights and possessions were in danger. The practical, more paranoid types realized that their lives were also threatened. Some emigrated elsewhere. Some stayed. Many who stayed would suffer the consequences of the Holocaust or spend years in labor camps. A few with names and social behavior not easily identified as Jewish converted to Catholicism and remained in Germany until the end of World War II.

Jason's family converted to Catholicism when Jason was four years old. Later he couldn't recall "Jewish experiences" prior to the conversion. It was a "total conversion." Physical evidence (books, letters, pictures, music scores, furniture) that might possibly tie the family to Judaism was destroyed. In and out of their home only German was spoken, never Yiddish. Relatives were informed not to write or contact the family under any circumstances. There was no discussion of the past in Jason's presence. He was not to know his religion of birth. Jason's parents spent considerable time at the local Catholic church. They participated in the activities of their parish and, at times, their diocese. Jason attended the local Catholic school. The majority of his playmates were from Catholic families. He went to their birthday parties and they came to his. They devised and carried out the usual childish ways of irritating the local priest, storekeepers, and their parents. His parents entertained other Catholic families and the local priests often visited their home. At meals, grace was said before eating food that was not kosher.

The 1930s passed, as did World War II. Jason and his family were never identified as Jews. If any of the people with whom they associated suspected they were, those suspicions were not revealed.

With World War II over, many of the Jewish families who had converted to Catholicism immigrated to other countries, some to the United States. It was the United States to which Jason and his family moved in the early 1950s.

The 1950s and 1960s were years in which Americans were highly receptive and sympathetic to Jews. The details and the death count of the Holocaust were becoming clear. Whole families had been disrupted, often killed. It was a time for America to be understanding, receptive, and supportive. And for the most part it was.

Jason's parents were delighted to be in America. For the

first time in two decades they felt physically and spiritually safe. It was at this point that they told Jason of his past. He had been born a Jew. They had converted to Catholicism and raised him as a Catholic to save his and their lives. Jason was now free to resume living as a Jew, as they planned to do.

Jason's parents had grown up in Jewish homes and they were able to revive much of their past and were again comfortable. Without conflict, they quietly separated from Catholicism and began their integration into the Jewish community.

For Jason, it was a very different story. Christmas, Easter, and Lent had become meaningful to him. He had celebrated them numerous times. High Mass inspired him and he went to great lengths to attend. He resonated to the music and the special, sweet smell of the ritual candles. He understood and laughed at Catholic jokes, the often voiced distaste by his friends for other religions, and their smugness about the implications of their baptisms and the specialness of being Catholic. He was comfortable in the environment of the local church, with its members, and in talking with priests about religion and other matters. Even though his parents no longer attended Catholic services, Jason often did, including taking confession at least once every other week.

Over time, he made several attempts to integrate into the Jewish community, to accept their values, their living styles, and to put his history as a Catholic behind him. He visited the local synagogue and talked with the rabbi about his past and his feelings. The rabbi and Jason's family were sympathetic to his affiliation with Catholicism and encouraged him to assimilate into the Jewish community gradually. He tried. But he never felt fully Jewish.

But there was another fact of which Jason became aware. A number of people in his community whose relatives had died at the hands of the Germans felt that Jason's family had somehow abandoned those Jews who died. They could understand the logic, perhaps even the necessity, of the conversion in order to survive. They could acknowledge that it had worked for Jason and his parents. After all, they were all still alive. They understood that Jason didn't make the decision to convert. They could even say that Jason was innocent and he should embrace his Jewish heritage. Still, they couldn't dismiss fully their thoughts that somehow Jason and his parents had survived at the expense of other Jews. Nor would their feelings go away. The local community was not fully open to Jason or his family.

Epilogue. Jason remained betwixt and between, pulled in two directions. He was no longer as Catholic as he was before being informed of his history. Yet he didn't feel like a Jew or view himself as one.

BELIEF AS PERSONAL TRADITION

Do these stories apply to only a very small percentage of people? For Jason, the answer is probably yes. For Elva and Robert, no. Are these stories of people who could be psychiatrically defined as delusional? Nothing in these stories or other details of their lives hints at such a conclusion. Wander into any religious event and count the number of people who suffer from delusions or some other mental disorder. The percentage may be lower than it is among nonbelievers, though the evidence one way or the other will always be controversial and may not even much matter.

Are these stories that affirm the truth of religious beliefs? Our answer is no. Belief—that which is held to be true—does not necessarily equate with truth. But for Robert and Elva, and to some degree for Jason, the answer is yes. For them belief is equated with truth. Moreover, there are simply too many similar examples of people who have found true meaning in their lives by embracing seriously a religion and its beliefs to dismiss such behavior as frivolous or meaningless. It's what many, if not the majority, of people do.

The stories are instructive in other ways. For example, they emphasize the influence of indoctrination. Jason's indoctrination into the Catholic Church would affect his life long after his parents revealed the past. There was part of him that remained Catholic and accepted its beliefs and history as meaningful to him personally. And this was despite the fact it was clear to Jason that he was not obligated by his Catholic experience. He could easily and readily believe differently than he did. So for Jason, the often-stated religious axiom that early indoctrination assures lifelong belief would seem to apply if only in part and especially given the extraordinary circumstances that influenced him.

On the other hand, Robert appears to have been immune to his early evangelical indoctrination. The axiom doesn't apply, but a reasonable assumption is that his early upbringing facilitated his rapid and full embrace of religion two decades later. With Elva we are uncertain if she was indoctrinated into the church or gravitated into Catholicism because of her association and love for Millie coupled with her vulnerability as a six-year-old who had lost her parents.

In our view, the most informative story here is about religious belief. What is noteworthy is its capacity to captivate the thinking and feelings of individuals. It can dominate their lives and provide them with satisfaction. It can certainly lead them to reject all other alternative beliefs. Logically, its power and command baffles.

Literally all reading this would agree if they had been born into and brought up in a distinctly different culture with a distinctly different religion than their own, they would be highly likely to think differently, be a member of a different religion, and look at other styles of life and religions with suspicion.[15]

Why then such deep and abiding commitment to one's religion and its beliefs? Something else is at work.

We've been proposing that the something else is to be found in the many things that religious and largely only religious beliefs provide. Earlier we discussed the fact that religious beliefs give answers to otherwise unanswerable questions. They provide a meaningful place in this world and perhaps the next. They elevate self-esteem and a sense of social efficacy, facilitate relationships, improve credibility among certain groups, and so forth. It is almost as if this something else in religion is a magic aerosol, which suffuses a wide environment with the clear certainty of a particular scent or aroma or remembered atmosphere. It is in that sense almost spiritual, because it is so ambient, so unspecific, and yet so palpable and even influential.

So this is part of our and others' answer. But there is also another factor illustrated in each of the stories: religious beliefs give meaning to the otherwise mundane and repetitive tasks of life that can easily become boring and tiresome. For example, taking out the garbage daily, trimming the hedge for the fiftieth time, and sweeping the driveway the one hundred and fiftieth time—these take on a different than usual meaning if one believes that the aroma of godliness is associated with order and cleanliness. Otherwise they are simply chores often to be avoided if possible. And from one perspective, why not? Why not view the mundane as helping create a Garden of Eden in one's own home? The idea is exciting, even motivating, that quotidian banal actions can cumulatively generate elements of remarkable moral grandeur.

We stress this point because the importance of religious beliefs is seldom viewed at the level of sweeping the driveway or trimming the hedge. But such beliefs permeate the truly committed daily. Religions may celebrate something as ordinary as the first of a particular seasonal fruit, blessing a modest dinner of tuna casserole, bringing old overcoats to the Salvation Army in the autumn, wiping the dishes dry, and feeding grateful pets a meal of food they like. An extra note is added to the chord of the music of daily life.

And why not?

Chapter 4

FAITH IN SEX

Three very different stories about different individuals under different circumstances made up the previous chapter. But each displayed the same life-defining feature— religion. Robert's and Elva's stories were about individuals who adopted religion and religion that adopted them. Robert found inspiration in the Bible and a supportive and convivial community of believers. Despite her personal losses, Elva found inner peace in her personal relationship with her God and the will to face adversity and uncertainty. The third story was Jason's, for whom religious conversion was driven by his parents' survival strategies, an unraveling of the past, and the emotional and cognitive tug-and-pull between early religious indoctrination and historical revision. This was coupled with the tumultuous feelings of an ethnic-religious community acutely aware of its recent history and the death of many of its members and relatives.

Here we turn to a far more general topic: faith in sex. Yes, faith in sex. Or we can put it as a question: why are many religions so seriously concerned and active in their attempts to guide, manage, judge, and adjudicate the sexual lives of their members?

We are especially interested in the endlessly gripping psychological and profoundly physical power of sexuality. We have to be. It intersects constantly—virtually from birth to death—with a vast array of rules based in religion that govern erotic and reproductive activity. Religion is with us every step of the sexual way. At the same time, sexuality is at the core of life and its renewal. In what purports to be a scientific document radiating cool dispassion, such as this one, it is difficult to render adequately the human turmoil these matters cause. An efficient guide to their nature is James Joyce's *Portrait of the Artist as a Young Man*, which could not be more urgent in its depiction of the archetypal collision between biology in the form of sex and religion in the form of Irish Catholicism.[1] Another scarcely temperate description of the landscape of this battle is Dante's *Divine Comedy*.[2] Both have nailed enduring announcements about *faith in sex* to the cathedral door.

Faith in sex. Could there be a broader topic? It seems unlikely if a comprehensive review was our agenda. But it is not. Our interest is in illustrating how the beliefs, values, and rules of religions bracket and punctuate the lives of their members during times of rapid social change.

THE CYCLE OF SEX

First some background.

Two huge issues—technical but huge nonetheless—define the issue. The first is that many of the modern ways of managing sexual behavior and reproduction are from a postreligious period of history. Decisive innovations such as the Pill, IUDs, ova and sperm implantation via mechanical devices, and chemical and mechanical abortions were developed long after many religions took it upon themselves to guide the sexual behavior of their members. All that our ancestors had to work with were the troublingly naked bodies of Adam and Eve. Second, literally all the various religious views and rules about sexual behavior were developed before the discovery of drugs that can cure many but obviously not all sex-related diseases. We can assume that as with dietary restrictions, many rules governing sexuality were medically benign efforts to restrict the hazards of intimate interaction. But with contemporary medical effectiveness, sexual enthusiasts are able to ask, is this cautionary trip necessary still? What's wrong with sex 24/7, with seven people or twenty-four?

Now to the cycle.

In a very accurate and descriptive sense, human life is created as a consequence of sexual intercourse between male and female. Time passes and a child is born. Both the child's and the parents' lives will be very much influenced by the birth of the child but also by its sex. For example, the child's sex directly influences the name or names he or she are likely to be given. How many males have the first name Joan or Midge? It will affect how he or she will be dressed by parents and what styles of clothing the offspring will prefer later in life. Sex will determine which manners parents will emphasize and which they will disregard or assign to low priority, what rules of social behavior they will teach and insist upon, what strategies they will authorize as acceptable in the pursuit of intimate relationships, and so forth. Sex has multiple meanings and multiple facets, which extend to important variants of the statistical normal, such as to homosexuals, transsexuals, and those who live lives of willful abstinence, especially for formal reasons, as among priests, nuns, and some monks.

One or more facets of sex are involved in each of these events and decisions. And as every parent and church authority knows, efforts to direct and at times control the sexual behavior of the young, to say nothing of everything else too, have only a partial chance of success.

Why? The answer: biology. Of course there are influences of parents, education in school, and the church. But sex has its own biotimetable. Learning about and experimenting with sex, whatever type, usually begins somewhere in the very early teens if not earlier and may continue in one form or another throughout life. Sexual urges may occur spontaneously as well as in response to obvious sexual and seductive stimuli (a fact that Hollywood, magazine publishers, and advertisers know very well). Moreover, there is only a slight chance that such behavior can be wholly constrained. On average, sex is fun, personally often spectacularly satisfying, self-defining, often novel, and often costless at the moment. The intensity of the need to satisfy such urges often leads to throwing caution to the swift wind. In turn, this can deliver unexpected and unwanted or wanted pregnancies and venereal disease. It can produce consequences for a person's reputation for prudence and sound personal management.

These points can be put another way. No religious official with any savvy expects their behavior rules to be followed 100 percent of the time. Rules about sex are no exception. Males and females are susceptible and often willingly succumb to well-crafted sexual messages. They are also the omnipresent source of countless and endless messages. Abstinence education has yet to be proven effective and many, if not most, teens in industrial societies have lost their virginity by age eighteen. Ever-earlier first menstruation has extended the time for female sexual activity and vulnerability. A tsunami of oral sex has roiled the high school and college-age communities. "Hooking up" and "friends with benefits" coyly describe forms of companionship that involve oral sex, if not full intercourse, but that preclude romantic ties or monogamy.

But an additional and drastically novel factor has also entered the picture: accurate DNA testing. In practice, among other things, this means that men can be decisively identified as the father of children to unmarried mothers seeking childcare and welfare benefits. States are only too happy to seek and gain judgments against boys and men who may face two decades of child support commandeered by a tough system. And the United States government even offers subsidies to states that succeed in garnishing or otherwise securing the wages of pleasure from males. All this high bureaucracy has at the margins enhanced the practical attraction of oral, not genital, sex, but it has not, of course, extinguished the causal behavior.

The Pill, of course, only lowered the threshold of opportunity. Efforts at reproductive guidance, such as by Planned Parenthood and various government agencies in different countries, have substantially reduced birth rates in communities that had once been seen as virtually uncontrollably fertile, such as Mexico, where two children is the increasingly common family size. It is now well known that population decline in various nations has become as pressing a socioeconomic and political concern as overpopulation was in the past. Many Catholic women use birth control pills especially after they have had several children. Members of other religions engage in sexual practices that reduce the chance of pregnancy significantly. More married women seek abortions than unmarried. Astonishingly, sterilization appears to be the most common form of contraception on the West Coast of the United States. This surely produces complexity for divorced people remarrying and perhaps wanting children with a new partner.

Still the cycle goes on, monthly, daily, nightly, in reverie and flurried action. Sexual urgencies appear and reappear and then appear and reappear again. There is ever-more candid public, and presumably private, discussion of sexual matters. The Internet has made available to every home with electricity anywhere in the world sexual images once restricted to tiny, red-lit neighborhoods of cosmopolitan cities. It has also enabled countless sexual candidates to seek and find partners, whose blandishments may include visual stimuli once restricted to the wedding night.

More customarily, with time, sexual activity becomes an integral part of marriages and long-standing relationships, although the manner of its expression and its intensity change.

Reproduction and the upbringing and management of the sexual development and behavior of one's children complete the cycle.

One way or another, all people are caught up somewhere in the cycle of interaction of rules and behavior, even if at times they denounce the rules and rebel against them. Paradoxically, rejection of the prevailing norms reveals how strenuous they actually are. Sex, and rules about sex, remain inescapable.

It is critical to recognize that each step in the cycle could be carried out independent of religion and without reference to the vast catalogue of theological comment on the subject. Surely this happened for millennia prior to formal religions, although presumably there were always beliefs about what was in effect a form of prereligious moral code. That such codes prevailed at one time and still do in a few remote locations is suggested by anthropological studies of groups with no formal religion. This we define here as the absence of belief in a higher or other-than-human power or influence and, very likely, an absence of written advisories on how to behave.

Nonetheless, these folks from the heart of darkness have clear codes about sex. For example, they will usually forbid sexual intercourse in public, usually sanction adultery within their group, establish parenthood and relationships of responsibility, and set a minimal acceptable age for sexual activity. When it comes to the management of the cycle, religions with rulebooks are latecomers.

A QUESTION AND ANSWERS

We have still to return to the giant pervasive question: why are many major religions so concerned and active in dealing with the various facets of sex?

Perhaps this chronic concern is the very source of their status as major religions. As strange as the question may seem, is managing sex tonight as important a feature of religious power as promising heaven tomorrow? The offer of sweet immortality is clearly a highly attractive product of the religious system—possibly its premier one, its loss-leader, its fabulous bargain. But is management of sexuality with all its often remarkable consequences also an important contribution of religious systems to their members?

This may be so, even if members do not find abrasive sexual inhibition tonight as agreeable as visions of luxurious afternoons tomorrow within the pearly gates.

There are a number of likely answers.

One is simple. People often need advisors about important or confusing elements of life. Religions are directly set up to provide precisely such advice and initially usually at very reasonable or no cost. That's what religions do, often very eagerly. It's scarcely a surprise that counsel may be most needed at the formal changes of status and intersections of life. Religions are able to show strength and provide *brainsoothing* precisely when individuals need it or at least think they do.

And with equal precision it is around these life-cycle events that religions generate rituals. These both channel but also celebrate the emotional turbulence that follows the practical social consequences of sexual activity. Therefore, religious entities become involved in responding to the birth of a child or the formalizing of a marriage. While there has been major erosion of the mandatory nature of such religious involvement, it was not long ago in France that the only acceptable legal names for French children were of saints, and in countless jurisdictions religious sanction was and is still necessary to solemnize marriage. Of course, a similar involvement of religious personnel occurs on death, which has nothing to do with sex except to announce its end here on earth. Yet the need for what we have called and will call in the future *brainsoothing* is obvious at such stunning moments. It is redundant to say that this emphasizes the similar role of religion in responding to the events of the sexual cycle.

Influencing the lifelong experience of families and individuals in the context of sexuality has forever been a specialty of religions. This is in contrast to other disciplines such as politics, urban design, economic planning, and even medicine. Religions have always shown and carried out a general tendency to manage and direct the lives of their members and their offspring. In effect, religions specialize in intruding, meddling in, supporting, protecting—what have you—the lives of the families with which they are associated. As we have seen, it is often difficult or at least unusual in much of the world for people to change their religious affiliation, which in effect supports their natal religions. And there is the sharply positive, if not essential, requirement that most parents are expected to raise their children within some religious structure.

Of course, many don't. We know from recent research that especially in a mobile society such as the United States, as many as 40 percent of Americans have left their first faith for another or for none at all. Some 16 percent of Americans claim no affiliation, with men more significantly likely to have no claim than women. And some 37 percent of marriages involve spouses of different religious flavors.[3]

Notwithstanding broader socioreligious patterns, Muslim parents are very firmly obligated to provide a strict Muslim environment for their children. Catholic parents are instructed to baptize their offspring. Jewish boys are expected to be circumcised, and both boys and girls are provided religious ceremonies at the onset of adolescence. A youngster's choice of marriage partner can be fiercely disturbing to parents concerned about religious lineage. While many of the most dramatic news stories may emerge from tight societies such as traditional Muslim and Hindu ones with such events as their occasional *honor killings*, the issue of mate choice and religion is rarely altogether casual.

Regardless of what happens in very mobile communities or at the margins, depending on which jurisdiction they are in, families may be required not only by custom but also by law to enroll their children in religious schools, and their taxation may be affected by that, too. Some entities, such as the Saudi government, may insert fundamental schools in otherwise alien communities for political and religious reasons of their own. For many religions there are strict rules for when, what, and where to eat and, of course, when, what, and where not to eat. There are rules for social relationships—for example, singles nights at Robert's church, where holding hands was the acceptable limit of physical intimacy, or with Elva, who was an unavailable prospect for male sexual satisfaction unless she agreed. Rules governing nudity, especially female and, of course, in public, have everywhere wielded firm influence over codes of dressing and undressing.

But what is new here? Isn't all of this obvious?

Yes. Precisely because this is so obvious we are obligated to acknowledge the pervasive power of the link between the religious and the sexual careers of individuals and their communities. As anthropologists tell their students, "The most important thing to know about a community is what it takes for granted." That there is a widespread and powerful trinity—sometimes even holy—of family, sex, and religion underlines the importance of the links we have been exploring here.

And religion may be a factor in enhancing sexual freedom

as well as constraining it. There are evidently signs even in strict Islamic society that interpretations of the Qur'an are provided to sanction not only female orgasm but also general oral/bodily contact. For example, a traditionally garbed sex therapist in Dubai has invoked the Qur'an to offer permission for oral sex while also achieving unexpected influence in a hitherto hidden world. A student who studied customs in rural Iran reported that public norms in rural households were strictly defined in the public rooms of a house. However, upstairs in the sleeping alcoves of the adults, *Playboy* Bunnies and similar representations of light sexual relief were permitted and abundant.[4]

Does this imply that there is nothing specific about sexual behavior in the minds of religious authorities and that it is just one of many behaviors requiring guidance? It cannot be dismissed this quickly. Most religions that concern themselves with families are relatively and often consciously conservative in thought and action. A focus on all aspects of the cycle may be *the* primary religious interest, not just improper sex in a bed in an assignation motel.

This is just part of the answer. The issue is not simply the control of sexual pleasure but also of its possible consequences. It was vital for people to enjoy confidence about the paternity of children before the emergence of DNA testing about paternity. "Who's your daddy?" was not merely a music-hall comedian's question but the core of endless drama, uncertainty, recriminations, and marital strife. Even today, geneticists estimate that some 10 to 15 percent of children are not the product of the men associated with them as fathers.

Obviously religious inhibition was one method of reducing unwanted pregnancies and their often undesirable or often at least challenging social consequences for both parents and offspring. This must be a substantial, if also painfully obvious, part of the reason for the preoccupation of many religions with female virginity at the time of marriage. The graphically lesser concern with male marital virginity suggests that the matter is less about morality and an embargo on pleasure than control of reproduction for social and usually religious ends.

Control of reproduction is an evident factor in the common goal of religion: carry the "Word of God" and increase the number of church members.[5] Reproduction is a means of adding to the roll call. This applies specifically to Catholics, for example, who are aware that in the New Testament there are several references to reproduction and training of offspring to serve Jesus and God. The demographic element here is surely part of the power of religion over social life, particularly where numbers of coreligionists have direct impact on ongoing politics, as in Israel, Iraq, Turkey, and francophone Canada. There are interesting exceptions to the "Go forth and multiply" groups, namely, those such as the bizarrely impractical celibate Shakers and certain members of other churches such as nuns, monks, and priests who are expected to relinquish active sexual lives to increase their spiritual purity and closeness to God.

And, come to think of it, why abstinence conduces to piety is not immediately apparent. It is in fact highly puzzling unless one assumes as we do that the management of guilt is an important coin of the religious realm.

The concern of religious authorities with sex is presumably related to the simple fact that sexual behavior is the most difficult of all human behaviors to control, short of eating, breathing, and excreting. The centuries-old and thriving business of prostitution suggests as much, and the explosive impact of Internet pornography is a recent demonstration of the same eternal principle.

If it is not directed, sexual behavior threatens a religion's authority. This is presumably one of the reasons why Catholic priests and nuns and certain Buddhist monks vow celibacy.

Worries about controlling the spread of disease must surely be a related factor. But even in communities with capable medical care, AIDS remains a defiant challenge. People will have sex even though there is clear evidence it may kill them.

As oxygen is to air, guilt is to religion. The management of what appears to be a human weakness to accept that, "Yes, O Lord, I am guilty," swirls in and among the personnel and behaviors associated with religion. We don't care to go into—not our job—either the theological folderol about the nature and the sources of original sin or the psychoanalytic proposals of the endogenous guilt that has been asserted to be an inevitable and volatile feature of the human psyche.

But guilt, and awareness of it, there is aplenty. An important product of religions is producing enveloping skill at reminding their members that they are or have been or will be guilty. Religions provide a mechanism for *brainsoothing* away the turmoil the brain endures: guilt and its cousin emotions.

Contemplate perhaps the most dramatic, if common, mechanism for accomplishing this *brainsoothe*: the Catholic confessional. Herein a guilty event or thought is announced by the citizen and an appropriate penalty is levied by the official. Until the following week, the cycle of guilty action is wiped away by redemptive action. This is truly a remarkable service Roman Catholicism provides and its ubiquity testifies to its involvement in something widespread among humans, often achingly real, and emotionally identifiable.

In other traditions, such as the evangelical, there is what we can think of as a massive *master confessional* in the form of nothing less than being born again, to which we have already referred. This is a behavioral event of considerable complexity. Suffice it to say here that it assumes a person has reflected seriously on his or her life so far. This life has been found to be wracked with guilty, destructive, and insupportable events. The person then concludes that a top-to-bottom personal renovation is desirable or even mandatory. This is likely to involve an embrace of religion and a god. And there is a well-tested procedure for greeting the newly sanctified former sinner. Born again, but without sexual activity by anyone. An amazing performance. It is not produced by the knee or the elbow or the kidney but by the brain.

For the truth of a higher authority to carry any influence, there have to be rules of behavior and rewards for carrying them out reasonably well. Consequences for not doing so are required. The confessional and rebaptism are revealing inasmuch as they show how otherwise banal events of normal guilty quotidian life—a roll in foreign hay—can be transformed into well-lit transitions to new and admirable personal status.

This reveals as well an effective mode of management of any large population of members. First, exploit the human biological tendency to incorporate rules and to act by them and then, second, for those who fail to follow the rules, lay on the guilt.

What a fascinating and exceedingly efficient solution for the administration of souls. Instill rules in their brains about what is for everyone an inevitable cycle of needful and potentially stormy sexuality. Then permit believers to act on their own and let those who violate the rules suffer guilt and seek repentance. Then, each week, encourage believers to voluntarily attend astonishingly costly cathedrals or humble shopfront chapels for a pleasing jolt of *brainsoothe*.

It is quite amazing. We have to confess we stand in awe of such an adept and effective means of managing the most passionate impulses of an endlessly gregarious and lusty primate.[6]

Chapter 5
RELIGION AS LAW AND THE DENIAL OF BIOLOGY

Histead istory is not just the past but is a living force. It rarely releases its grip overnight and can insinuate itself into the present, often tenaciously. Entire communities may be highly wary of relying on secular rather than sacred authority to supervise social behavior—would one rely solely on oneself and some chums? This despite the fact that people act much the same way everywhere, which suggests (or should) that much of their behavior is rooted in biology.

A dramatic and symbolic case is demonstrated by what happened atop St. Basil's Cathedral in Moscow's Red Square. This is the ceremonial historic heart of Russian society. For centuries there had been a crucifix atop the steeple. But what to do after the Russian Revolution, which was to be secular and unafraid of challenging the certainties of the past? In fact it took a full generation, twenty years, for the religious crucifix to be replaced by the secular Red Star. Only in 1937 was there sufficient confidence in both secular and political authority to make such a significant change.

Was the confidence truly sufficient? Only among a few it

seems. The end of the cold war served as the entrée to a resurgence of the Russian Orthodox Church and multiple forms of religiosity so long officially suppressed.

One idea fondly and understandably advanced by religious adherents is that obedience to a higher power is essential to social order and general morality. Without the presence and the influence of religion, daily conduct surely would be absent of order and meaning. The companion idea is that religion alone can do the job properly and thoroughly.

There may be a variety of objections to these views.[1] But they are plausible and powerful assertions given the ubiquitous power of religions over personal behavior and decorum.

Let's consider. If some 80 percent of the world's adult population believes that there are extrahuman sources of creation of human behavior; if there is no convincing scientific evidence supporting these beliefs; if there is compelling biological evidence that the source, as well as the object, of religious beliefs, feelings, and behavior is the brain and that they serve to *brainsoothe*; and if the majority of believers reject or disregard the biological evidence; there must be reasons why.

This chapter is about these reasons. It is about how religious beliefs and feelings function as rules of behavior and why the majority of believers prefer religious explanations of these rules to biological explanations.

Where to begin? As we discuss in detail in chapter six, an obvious place to look is among our closest primate relatives—chimpanzees. If they have social order and behavior consistent with what humans consider general morality, then we may have found a point of departure.

As a matter of now-amply recorded fact, chimpanzees comport themselves in a manner suggesting that there is some consensual moral system they share, whether or not there is one. They have rules. Their behavior is predictable and punishments are doled out for rule violation or indifference.[2]

Humans of course are far more articulate and probing. And for chimpanzees we have to search keenly for what, if anything, may substitute for religious belief as a reliable and durable cement of social structure.

This is not a new concern but one of deep age. In the past, an unanticipated and subsidiary problem of societies seeking a secular basis for social life was basic: where do rules of behavior that are not formally legislated—*formal laws*—gain their justification and authority?[3] These may be called *informal rules*, or *conventions*, and they include social protocol, relationships with neighbors, how to deal with relatives and offspring, how to behave at public and private events, how much effort to contribute to the public good, how quiet to be in a library, when it is acceptable for men in cities to wear shorts, and so on. This is a concern as yet unsolved.

In certain societies, such as France and England during the 1800s and 1900s and the United States between 1780 and 1860, people responsible for legislating formal laws in parliaments and other assemblies assumed that informal rules should and did have their origin and justification in religion. (The exceptions were specific local rules; for example, "Rattlesnakes Live in Woodpiles" is a relevant warning for the mountains of California but not the seaside and, "Beware of the Undertow" is a relevant sign at the shoreline but not for a barely trickling mountain stream.) And it turned out that the formal legislators were largely correct. Religions had long histories of supporting informal rules or at least accommodating them if they didn't challenge legal values.

A fair amount, let's say 20 percent of human behavior, is either strictly legal or illegal from the perspective of formal law. Laws dealing with driving, theft, larceny, marriage, divorce, property, taxes, investments, rape, and child molestation are familiar examples. These reflect what those who manage society believe should apply to all of its members with only very careful and specified exceptions.

Many of these laws can be traced to religions and their sacred texts—both the Qur'an and the Old Testament have a good deal to say about theft, marriage, property, and how and what to eat and when. However, it is strikingly clear from observing other animal societies that departments of theology, an ark, an altar of sacred texts, or indoctrination are not essential for species to develop rules of behavior. The rule, not the exception, is: rules are followed. This behavior blends the formal and informal in different species. Remotely located chimpanzee groups isolated from each other in the African forest share much behavior in common. But within limits they also have clearly different local cultures.[4] This suggests that they adopt a portion of their behavior to local conditions and even to their own cultural traditions. Moreover, this point applies to literally all well-studied animal species. Birds, carnivores, fish, and reptiles all have rules by which they operate. Even bats in caves can recognize members of their kin group and treat them preferentially.

There is a clear implication of this: biology is the source of many of their rules. Are humans different?

Again, let's consider. For secular societies and using our arbitrary 20 percent rule, this leaves perhaps 80 percent of daily human behavior that is not specifically legislated. Such behavior can be full-bore strict—you had better stand when you salute the flag and sing the national anthem and bow your head when a minister prays at a funeral. It is much less obligatory to help elderly folks carry packages. But often within rather surprisingly narrow limits, rules or conventions differ very little within a given society. A person brought up in Nevada will find little trouble adapting to life in Maine. The velocity of American migration from state to state and region to region reveals the potency of often tacit but usually knowable or rapidly apprehended conventions.[5]

Often such rules may appear to originate in religions and

sacred texts. But nothing in the Bible asserts that one must kneel before entering a pew or bow one's head during prayer. People just do it. So clearly, for both chimpanzees and humans, sacred texts are not required for the development of rules, although established religious practices often are a source.

Still, the many advocates, critics, apologists, and enemies of secular societies have almost totally turned away from the proposal that much of the story of the 80 percent of informal lawful behavior can be explained by biology. There are exceptions such as Graham Wallas and William James, who over a century ago looked to biology for rules and norms as well as for explanations.[6] And a number of recently published books have seriously addressed the biological question.[7] In contrast, Karl Marx was said to hope that his work, which contained no biology, would provide laws for economics as firm as those Darwin provided for animals. He even sent a copy of *Das Kapital* to Darwin, which was politely acknowledged although apparently never read. Did his system falter and then fail because it ignored biology? Did communists get human nature wrong? And it remains evident from the writings, utterances, and street-corner pamphlets produced in dozens of nonsecular theocratic societies that the biological bases of behavior remain largely irrelevant. They are often, indeed, deemed not only inapplicable but also fierce threats to the sanctity of whichever perfect and revealed divine scheme is being protected.

Yet if birds, fish, reptiles, all known nonhuman primate species, and human beings all have informal rules, isn't biology the best place to look for their source?

Two moments inch us toward an answer to this question.

TWO MOMENTS

Chief Fred

Author McGuire relates the following experience.

I was in West Africa searching for a species of monkeys. We would travel by water after the roads ended. It be faster than chopping through the jungle. Two guides were hired. A canoe was rented. The three of us started upstream. Days 1 and 2 of the trip were uneventful and unproductive. No nonhuman primates. On Day 3, we arrived at a small village.

The villagers—some of the children had never seen a "white man"—were friendly but cautious. We disembarked. Villagers escorted us to an open space in the center of a group of huts. The chief of the village—I will call him "Chief Fred" as this was close to the way his name was pronounced—presented me with a bowl of soup containing moving worms and buglike organisms.

One of the guides informed me: "You are expected to drink the soup." I drank the soup as the villagers watched. Fortunately, I had Tums in my pack and quickly popped three into my mouth. When the soup was finished, the guides and I were accepted.

Time passed. The last rays of the sun departed from the forest. We were shown a place to sleep for the night. Then came supper or its local equivalent and the talk and chatter that accompanies meals (and Tums).

One guide was a natural linguist, fluent in the local dialect and English. I asked the chief why they had developed the "soup custom" for visitors?

"Those who drink the soup do not hurt us."

Supper continued and discussion eventually turned to the source of the rules of behavior such as when and why villagers share their food, who is responsible for children, and how breaking rules was handled.

"Years ago, a missionary came to live with us. He advised us about how to behave and how to respect and worship his God. He promised that if we behaved the way his God wanted, we would be rewarded in this life and after. Our children would flourish, our food supply—it was often uncertain—would improve and stabilize, and we would find an inner peace. We believed his teachings and became Christians. We were baptized."

"Have events unfolded as the missionary said?" I asked.

"No. We soon encountered several years of famine. In another year, one-third of the village died because of disease. We ushered the missionary out of the village."

"What's happened since?"

"We have devised our own rules. Children, mothers, and elders are fed first. Parents are responsible for their children. When we need to build a new hut, everyone participates. Those who don't cooperate are sent away. We are now healthier, eat better, and there has been very little disease."

"What happened to the missionary's God?"

"Some of the elders still believe in him—the missionary was here fifteen years."

"Do you have another god?"

"No . . . or maybe yes. Our god is our ability to survive and remain healthy. This we have solved for ourselves. The missionary's God is not for us."[8]

The Ladies Jane

Author McGuire is at a ranch located in the tree-covered rolling hills of the western Sierra Nevada. Much of the ranch is visually inaccessible from the ranch house. Cows range over most of the acreage. Each morning someone "rides fence" to check that the fences are intact and to identify sick animals.

I am riding fence the day this moment begins. It is winter and cold. An hour's inspection reveals no problems with the fences. The cows are healthy and active. As I start the ride back to the ranch house, I notice a thin stream of smoke coming from a far corner of the ranch. I ride over to investigate. There I meet the Ladies Jane, three females in their late teens or early twenties who have a tent and are attempting to cook a dead turkey (likely the product of poachers) with a fire barely sufficient to warm a sausage. I dismount and our discussion begins.

"Hello. I am Michael McGuire and I live here. . . . Are you aware this is private property?"

To my surprise they acknowledge the point.

"Then can you tell me why you are here?"

"Our minister advised us to seek our salvation."

Each had a checkered social and legal history. They had embarrassed their families. They had been shunned by many of their friends. And they had decided to change their ways. They would "go to the wilderness . . . live alone . . . atone for their sins . . . and decide how to live their lives."

Their story seems genuine. There are no obvious deceptions, no flirting, no request for my assistance or my evaluation of what they were about. But it is January. The temperature at night often drops below 32 degrees. Their tent is flimsy, and (through my eyes as a doctor) one of the ladies appears to have pneumonia and another appears severely undernourished.

I return to the ranch house, prepare a hearty meal, round up some blankets, and return to their campsite. They eat as if they hadn't done so in several days. They thank me.

"I will return in the morning and we can discuss what to do."

They're eager to see me the following morning, more eager still to eat the breakfast I brought them. We talk.

"Is your effort to atone making any progress?" I ask.

"We haven't decided."

For a time we talk about what is required when one aspires to atone for one's sins. It is clear that they are bright and sane young women. There is nothing in their behavior suggesting that they are mentally ill or taking drugs. What is clear was that they are believers—their God will lead them to full atonement.

Eventually the conversation returns to trespassing.

"You realize that what you are doing is against the law and that you will have to leave the property soon."

"Yours is not a law we recognize," replies the youngest of the three. "We recognize the law of God and the Church. Your laws don't recognize our God. We do. We will stay until our atonement is complete."

I left them and visited their minister. He came to the ranch, spoke with the women, and, as a group, they departed. Months later I inquired of the minister about their fate. "They are on the right track now. You were very helpful. I thank you. God does too."

*   *   *

What is to be made of these moments?

Chief Fred was a biologist in the rough. The facts that he valued were those of survival, health, and the well-being of his village. It was these facts that he and the villagers used to develop their formal and informal rules. The addition of a new god to their rule book offered no obvious advantage, especially in view of their unsatisfying experience with the now-departed missionary's god. Their rules were firmly grounded in the realities of their environment and their biology—eating, surviving, and helping one another. For them these required no obedience to a higher authority, no sacred texts, no interpreters of such texts.

For the Ladies, Jane rules came from the *top down*. The rules of their God superseded formal rules and ignored the possibility of a biological basis of informal rules. Both were largely irrelevant to what they valued and how they planned to live their lives. Living by their rules led to a dangerous denial of their biology. Two of the ladies were ill, one severely under-nourished, another with pneumonia. Both required medical treatment after departing from the ranch. Their tent was insufficient to protect them from the cold and the wind. The fire they had built would have taken a week, if not longer, to cook the turkey. They were unprepared for the practical living requirements of their journey of atonement. Yet none of this seemed relevant to them. For them, the only powerful truth was "God will lead us to full atonement."

A BIOLOGY OF RULES—YES OR NO?

As the story goes, Sir Richard and Lady Burton were exploring together. With each new culture they encountered, Sir Richard perceived primarily its difference from other cultures while Lady Burton saw its similarities. Differences certainly are present. Compare the daily life of a Bushman in South Africa with a same-age, same-sex stockbroker working on Wall Street or a traffic policeman directing rush-hour traffic in Singapore. They dress differently. They speak differently. They act differently. They eat different things. The differences clearly seem to outweigh the similarities. Behavior, dress, food, and language in Paris, Texas, isn't what it is in Paris, France.

But a closer look at different groups and cultures suggests another answer: yes, there are differences, but there are similarities, too. Lady Burton was also right. People of literally all cultures share many behavioral basics. Look beyond the differences in language, food, dress, and appearance, and people everywhere do much the same thing. They engage in lasting relationships, reproduce and care for their offspring, treat kin differently from nonkin, and form groups and make or adopt rules. This was the case for Chief Fred and the villagers and the Ladies Jane as well. And like Richard Burton, they tended to view other groups as different from their own.

Similarities don't occur at random and there is no compelling evidence of a worldwide conspiracy for everyone to act the same way. Mothers and fathers don't have to learn or be taught to love and take care of their infants. Despite the often quarrelsome hours during childhood, siblings don't have to learn that they have certain responsibilities toward each other. Teenagers don't have to learn to have sexual feelings. Adult males don't have to learn that it is necessary to compete over resources in the marketplace if they want their share. Adult females don't have to learn that having offspring by some males rather than others bodes better for the quality of the lives and the future options of their offspring.

Or take a behavior as ubiquitous as reciprocity. *A* may be a poor hunter and *B* an excellent one. They work together: *B* does the hunting and after the kill, *A* skins, cleans, and prepares the animal for eating. This was the rule for Chief Fred and the villagers. Both *A* and *B* get the meat they desire, both save time. It works the same way in the city. *A* is a good automobile mechanic and helps *B* fix his car. *B* is an accomplished carpenter and helps *A* repair his porch. And of course if *B* doesn't repay *A*'s help, he will be typed as a nonplayer and experience the social consequences that accompany the label. It's such informal rules that structure and guide the behavior of the *A*s and the *B*s of the world. They occur billions of times each day. Reciprocity is like breathing. It is at the center of society's heartbeat.

For a group of today that aspires still to be a group of tomorrow, it is critical that these rules function very much as if they were formal rules. And usually they do just that. The vast majority of favors are repaid. Reputations are made or destroyed by following or disregarding rules. Rule violation leads to real punishment designed primarily to reduce behavior rooted in biology.[9] If a person fails to support his kin as expected, kin usually marginalize him. His social reputation will wither. If a mother fails to care for her offspring, she may be ostracized socially. Moreover, responses to violators of informal rules may be more consequential than if a formal law has been broken. Social ostracism for petty theft among a group of chums or a closely knit family is often more severe and lasting than what the formal system hands out for similar behavior, such as a fine and several months on probation. These are the biological rules of social behavior as distinct from secular rules.

Yet, amazingly, different groups explain these behavioral basics and similarities very differently. This despite the fact that there is a very extensive and robust literature that describes the biological bases of why behavior is organized and prioritized and not chaotic. This has emerged from looking at the nature of genes, physiological systems, and social structure. Biological theories and explanations deal with an array of profound issues, such as the source of self-interest and its many behavioral facets. We now know much more about why groups organize and why individual members suppress self-interest because of possible other benefits they may derive from participating as rule-following group members—reciprocity, for example.[10]

There is also a substantial knowledge about why people invest preferentially in their kin compared to nonkin and why people become committed to the groups in which they are members and to which they devote resources. There is good writing and thinking about the role of deception—a rule violation even among thieves—in interaction and on self-deception regarding how people evaluate their relations with others. Such work does more than go a long way to explain how people treat their friends, their kin, and those who do or might harm their kin or friends. It provides a basis for explaining the similarity of these behaviors across vastly different cultures. Chaos is not the default side of human societies.

This is exactly what we see and are finding out daily about other species. Chimpanzees have relationships between kin that differ from those between nonkin. Animals enjoy group affiliations, which develop at some cost to their immediate self-interest but at some possible gain to broader features of their survival and well-being. It's irreducibly true that survival probability is greater among group members compared to loners.

Nonetheless, it is highly revealing that people of sharply different and often openly conflicting religious and ideological affiliations—let's say Christian evangelists and atheists or Hindus and Islamists—treat their kin, their friends, their neighbors, and their enemies in highly similar ways, so too with their vegetation, exterior décor, and mail delivery. It is also highly revealing that such behavioral basics acquire the status of social rules grounded in social nature, which is largely how they are explained, while other possible explanations are ignored. This is reality. That it is is thoroughly and even decisively pertinent to answering questions about the origin and theological status of morality, as well as formal laws and informal rules. The behavioral similarities across a thousand cultures and thousands of religions invite, if not demand, the idea that biology is the foundation and structure of such behavior, despite what differences there might be in their outward décor—differences that anthropologist Robin Fox has aptly called "ethnographic dazzle."

A FORK IN THE ROAD

For believers, a key issue—often, a rock-solid commitment—concerns establishing informal rules and, where possible, formal laws consistent with their religious beliefs.

This commitment suggests a sharp difference between one kind of analysis and another. Are we talking about split brains? Improbable. Are we talking about the pervasive power of belief and its capacity to insinuate itself into literally every idea and behavior except perhaps for the most mundane, such as a bout of indigestion after eating a bowl of soup full of worms and bugs followed by Tums? Very probable. In short, in this view, at most every level of daily life and interaction, religion reigns supreme over how the majority of people develop and execute reasonable informal rules for living. And they deem these consistent with the story of Adam and Eve, the idea of original sin, or what Allah prefers, what their private god approves of or expects.

There are exceptions. Those who deny biology at the grand level would likely acknowledge it if they were suffering from a broken leg, a body covered with poison oak, or an ingrown toenail associated with throbbing pain. Even for them, there are some things that their creator doesn't control. This was the case for the Ladies Jane when they accepted medical care. But the number of things their creator doesn't control is very few.

It is precisely these differences that are the nub of the religion-as-law and the denial-of-biology issue. Religions of all types, and fundamentalist religions in particular, support a far greater number of religion-based informal rules than nonreligious individuals or groups. A far larger percentage of daily behavior is prescribed and subject to the guidance and judgment of one's religion and its members and managers. The 20-80 percent ratio discussed earlier can approach 80-20 or 90-10 percent. In other words, 80 to 90 percent of behavior would be governed by religion-based rules in such fundamentalist religions.

Why have things turned out this way? There are answers.

Over history there have been numerous attempts to devise informal rules independent of existing religions, for example, communism, existentialism, and some secularists. On balance, these have been far less successful than hoped for initially. We suggest this is largely because they lacked ways of developing consensually validated informal rules. Why?

The percentages of informal rules that are often large do not unambiguously and agreeably have their source in religion. As a result, secular societies inevitably and presumably unintentionally experience increased conflict among members. Public objections and legal tactics aimed at removing religious symbols located in public places in the United States are familiar examples. But this is only one facet of the story. Education (what should be taught) is another. Public school curricula function as informal rule indoctrination for those who attend school. Social conventions are still another. Alcohol consumption and business hours in certain regions are examples of disagreement between sacred and secular rules. Personal preferences also enter the picture: who decided and when that bikinis shouldn't be worn to high school? And there are the well-known, well-advertised, and unresolved conflicts over abortion, stem-cell research, gay rights, women's rights, grandparent's rights, and on and on.

It's as if secularism invited the idea that every person has the right—perhaps even the responsibility—to devise and live by his or her own informal rules. This works for living alone in a remote mountain shack, but it becomes a demanding strategy if one wishes to get along with one's neighbors with often sharply different notions of what's right and good to do.

Moreover, secularism is not cheap. The conflict and cognitive and emotional disagreements it invites *are physiologically costly*. This is a central theme in our formulation that deserves repeating. Conflict is physiologically expensive and personally and socially aversive. Its consequence has many names, but *stress* is the most common. It is physiologically costly to try to intuit the motives of an opponent, to spend time defending against disagreement. It is costly to try compromise when an opponent disagrees totally with what you believe, to navigate daily through a potentially hostile world, and to suffer disappointment repeatedly. Why not settle for a less complex and more predictable system?

Most religious individuals, however, believe in a higher power. The authority they attribute to that power exerts an amazingly strong grip on how they think, feel, and behave. Given this authority, from their inception, many religions have taken the position that only they are responsible for the development, interpretation, and enforcement of informal rules.

In many ways, religions have been effective even in secular societies in establishing themselves as the rightful source and arbitrator of these rules. This is not surprising. Usually their rules are easily understood and easy to carry out. The Ten Commandments are readily understandable, reasonable to live by, and it's straightforward to expect one's neighbors to do so also. Those of the Qur'an are equally understandable and reasonable.

Such rules are already there and available to each new generation. Those who don't object strongly to their source don't have to devise rules for themselves. And if others are already following them and there are consequences for not doing so, this promises predictability in social interaction, which people seek and which is so central to social order.

It is possible in practice and in science to argue that biology offers an even greater authority than religion. Humans are a product of their biology. That seems certain. And for some individuals who claim their acceptance of biology does not interfere with their belief in a higher power, granting authority to biology or one of its products, the brain, is quite acceptable. Logical and provable, yes. But exciting and inspiring, perhaps not.

And what is behind all this? For the majority of humans it appears that biology lacks critical and highly desirable features compared to those offered by most of the world's religions. There is no afterlife except for mindless genetic replication subject to the whims of a swirling soupy natural selection. There is no readily available list of behaviors and feelings that, if followed, assure social approval and self-respect and possibly a privileged passage to life after medical death. What biology seems to do is strip away these features and provide far from inspiring details about the influence of genes on behavior, the chemical details of the brain, and identify those parts of the brain that are active during religious moments.

Small wonder religions take the cake.

Chapter 6

IS RELIGION
MONKEY BUSINESS?

We watch them quietly from a respectful and prudent distance. A group of chimpanzees has completed its morning forage for food. Now it has gathered in a quiet and soft assembly. The group's leader is present. But he performs none of his usual brisk managerial actions. He appears content to reassure and enfold individuals in the larger social context. The clearing in which they lounge is shaded and enfolded in protective vegetation. The individuals seem so intrinsic to the group that were they humans, they could be wearing T-shirts or other formal insignia.

The bonds of society of this particular group of beings are worn lightly but confidently. If a physical test could be contrived to display their social links, thick ribbons of burgundy fabric would be seen to tie them together. Some quiet rumble from a pipe organ or some gentle hymns would seem quite in emotional order. No evident system of symbolic belief cradles the group. No belief, that is, other than that these are their only lives and the only group in which they are living them.

They may not need a system of symbolic belief because the swirl of social reality is sturdy enough.

Whatever one's view on religion, God, an afterlife, or the origins of the universe, it often seems strikingly obvious that humans are a species apart.[1] A snake slithering through the grass; a hummingbird suspended in midair, defying gravity; a dolphin propelling itself eight feet out of the water—where is the humanness? Yet when watching chimpanzees and Old World monkeys, it's easy, if not inevitable, and even over-whelming to see resemblances to family, friends, and acquaintances. The similarity is simply too close to overlook, a point captured by one of the variations children sing to their friends at birthday time:

> Happy birthday, dear Billy,
> Happy birthday to you,
> You look like a chimpanzee,
> You act like one too.

In this chapter we ask: does religious behavior among humans build on a brain-body scaffolding shared with chimpanzees? That is, do they share genetic, physiologic, and behavioral features to the degree that strong inferences are permitted about common origins and functions of behavior? If the answer is yes, as we think it is, it invites a rigorous reappraisal of the origins, uniqueness, meaning, and function of religion. We will explore important features of chimpanzee life, which provide them forms of respite, solidarity, pleasure, and purpose. In a persuasively haunting manner, these resemble the impact of religion and spirituality on human society.

This isn't the first time this type of question has been asked. A number of scientists have suggested that chimpanzees have, understand, and behave by rules of right and wrong.[2] For example, they clearly grieve for deceased chimpanzees once close to them, sometimes with tragic persistence. They appear to treat fellow chimpanzees with physical disability in a supportive and nonexploitative manner.[3] They teach each other ways of securing and preparing food, such as termites in fallen trees and food that has to be cleaned of sand in water.[4] And who can forget the Jane Goodall film of a chimpanzee group permitting a female and her infant to join the group after she had respectfully and appealingly hovered on its outskirts for a period of time?

No cautious film director or theatrical impresario would dare show the melodramatic scene in which she makes a tour of the circle of the group and holds out her hand to each member that is grasped in return. Now she is in. But evidently she had first to acknowledge that she had once been out and could be so again.

The chimpanzees had laws about immigration. They appear to have others too. Rules of right and wrong and who is in and who is out are standard pillars on which religions build. A careful look at the species most closely related to humans should provide insights into the origins and functions of religion among humans. At the very least, how chimpanzees handle issues we would call "moral" conveys a powerful emotional jolt of recognition of what they are doing.

INNER ENVIRONMENTALISM

In a first obtuse but finally definable way, this becomes or should become a feature of the environmental movement, which has been mountingly concerned with nature and its preservation. This is the nature around us. But there is also a nature within us. Just as despoiling the human environment also affects the welfare of chimpanzees and other animals, so there is also an inner behavioral environment to be protected or trashed. And there may be clues about preserving our own inner environment by inspecting the one that chimpanzees unfold to us with increasing clarity and range. How they approach and solve moral issues is of sharp inner-environmental significance.

And these are not canaries in the mine—they are our cousins in the neighborhood.

Decent scientists have shown great caution about comparisons between humans and other animals and even chimpanzees. They plan to wait and see. In part, their reservations stem from the fact that the scaffolding question is problematic, not only because it is a big issue. But it immediately encounters potential analytic and interpretative difficulties.

For example, one possible answer could focus on the assumed common ancestor of chimpanzees and humans. If so, not only would this be highly speculative, but it would also rely on a long-extinct species thought to have lived no less than between six and eleven million years ago.

That seems shaky indeed. But can a sounder answer come from what we know about present-day chimpanzees and humans?

Yes. Far more is known about chimpanzees and humans who are alive than about their now-deceased ancestors, for which there are only scattered remains. Less obvious is the fact that if the inquiry is based on living species, it adds a special challenge: religion, at least formal religion as practiced by humans, appears to be no more than seventy thousand years old. This estimate is based on the dating of rock art depicting religiouslike behavior and stone carvings that suggest an appreciation and awe toward a serpent or comparable deity.[5]

When making estimates about past events, it's wise to allow leeway. For example, currently the oldest known skull attributed to *Homo sapiens sapiens* is approximately one hundred sixty thousand years old.[6] But older skulls with similar features are likely to be discovered. Thus, seventy thousand years might best be thought of as seventy thousand plus, say, another seventy thousand years.

If these dates are approximately accurate, there are at least two formidable implications. The first is that sometime in the last one hundred forty thousand to seventy thousand years, a combination of evolutionary events and circumstances led to religion. And the second is that if over this time period chimpanzees have evolved and changed less than humans (this is the current consensus among scientists), this difference could facilitate an identification of features of the shared scaffolding we have described. Necessarily, it will also stimulate questions about those unique and characteristically human elements that have been added during the last thousand or so centuries.

A further point. Scientific concepts are continually subject to revision as a result of new discoveries. This is particularly so for literally all the topics discussed in this chapter. They are hot research areas. Among other things, this means that there are more than the average number of uncertainties and speculations. So some of what is said here will and should—should! —undergo revision. Nonetheless, the idea that humans and chimpanzees share a common ancestor, that they diverged from this common ancestor very long ago, and yet that they have many commonalities has attracted rock-solid scientific support for over fifty years. This alone suggests that an inquiry into the commonalities might be highly informative.

SHORT BRIDGE OR WIDE CANYON?

We turn to the similarities and differences between chimpanzees and humans. For now, the discussion is general. But be aware that we will focus soon on similarities connected to religion and behavior like it.

Chimpanzees and humans look like each other and we like

looking at each other. If chimpanzees bothered to create zoos, almost certainly humans would be among the primary and most popular exhibits. We have dexterous hands. And, like the chimpanzees, we use separate hands or at times both hands for different tasks such as picking fruits, grooming, signaling, or smashing nuts with rocks.

Inside the skin the story is much the same and even clearer. It would be quite possible to teach human anatomy to medical students—the pancreas, the kidneys, the knee bone, the heart, and the lungs are close to identical—and brain surgery to aspiring neurosurgeons using chimpanzees.

Yet there are differences in the brain. So too with the muscular and anatomical apparatus for speech. Even if chimpanzees wanted to speak as humans do, they just can't.[7]

Body language is used for signaling social messages such as submission, affection, and intimidation.[8] The highly expressive faces of chimpanzees reveal and communicate emotions ranging from empathy and affection to hate and anger, traits that Hollywood exploits with its chimpanzee stars.[9] And both species know who they are. We humans are an inward-looking species with countless mirrors in the furniture of our lives. But chimpanzees also recognize themselves in the magic glass.

Are both specimens arrested by themselves in the same way and do they see something similar?

And similar development patterns are present. Both flavors of parent want to and do care for their infants and seem predisposed to teach. Offspring are born with a predisposition to learn from their parents. Like human children, they rather rapidly inhale the community's norms of conduct. These often include sophisticated behavioral rules and cooperation extending beyond two-animal or two-person altruism.[10] That is to say, they seem primed for cooperation for which there appears to be no apparent reward of frank self-interest.

Both species understand when it is wise to adopt certain

behaviors. For example, among chimpanzees, if there are alternate ways of solving the same problem, the solution favored by the group's dominant male is usually adopted.[11]

Might they understand that imitation can be a compliment? And even a complement?

And both species save tools for future use for obtaining food or hunting. Gorillas as well as several monkey species engage in similar behavior.[12]

Both form sounds for communication. While a complete chimpanzee vocabulary remains to be described, there are hints of similarities with human vocalization at moments of emotion and in sounds emitted during play. These may even be ancestral forms of normal human laughter. And we both can communicate using sign language within our species groups and between them.

But there are also obvious differences in language capacities. For example, there are computational constraints that limit not only chimpanzees but all nonhuman primates, such as the capacity to generate a near limitless range of meaningful verbal expressions, of which most humans are capable. But both can perform complex cognitive tasks, such as a group of chimpanzees coordinating its behavior when it raids or evades another group.

At times chimpanzees exhibit sensitivity to the plight of others. There are reports of chimpanzees that drown in the moats of zoos in efforts to save members of their group. Yet at other times they are indifferent to the plight of nonkin group members.[13] Both species cause pain through ridicule, shunning, and ostracism, behavior that has persuaded some authors to draw a connection to recurrent behavior among religions such as excommunication. And when there are conflicts, postconflict resolution or making up often follows. Among chimpanzees, this primarily involves reciprocal grooming, which quite clearly produces a social comfort zone for the animals involved.[14]

HIERARCHIES OF THE SPIRIT

Both species display hierarchies and both develop them spontaneously. For example, human adolescent males and females previously unknown to each other who are thrust together in living quarters when they attend summer camps make the formation of an accepted hierarchy a first order of business.[15] Introduced into similar conditions, unfamiliar chimpanzees have the same first social priority. Perhaps more than humans, chimpanzees divide and rule. Usually a dominant male rules an entire group often with the assistance of allies. Yet, much like humans, in any group, there are separate male and female hierarchies as well as kinship or family hierarchies that don't track the formal political ones and may undermine or challenge them.

Culture means social learning, the transfer of tradition, and developing the capacity to handle social complexity. This, too, humans share with chimpanzees. Chimpanzees that migrate from other groups—usually females—are taught the rules of their new group by adult members of their new group. And, as with humans, there is a conformity bias among chimpanzees. Different groups act differently, much like humans who live in different social environments. When in Rome or the Gombe Game Reserve . . .

Both species can be and often are aggressive, a feature characteristic of much of the primate line. Some investigators have suggested humans and chimpanzees share genes or a genetic array for aggression. Both prey on members of other species as well as those of their own, at times to the point of murder. Chimpanzee males rip off the testicles of their victims just as Nazi camp guards did and gangsters do still.

Among chimpanzees, such behavior is often contrasted with the popular and pleasing behavior of bonobos, a closely related species once notable for its lack of aggression and the presence of sexually and politically dominant females.

Bonobos enthusiastically employ vigorous sexuality as a way of reducing social tensions and sustaining a wide array of relationships.[16] This has of course endeared them to many commentators and wishful observers. But compared with chimpanzees there are very few bonobos alive in the wild. This provokes the dolorous conclusion that however appealing is their sensual conviviality, their lack of alert and snappy aggressive behavior may render them vulnerable to predation. Furthermore, research on wild bonobos reveals that both males and females hunt, that fighting is rampant, and sex in the wild differs from hook-ups in dormitories or zoos.[17] As we shall see later, the bonobo sex style is unlikely to find acceptance among the moral managers who find employment and power among the world's major religions.

Many of these similarities are not unique to chimpanzees or humans. For example, a number of bird species, to say nothing of squirrels, have impressive memories regarding where they have stored food—a form of planning for the future—and show clear signs of intelligence.[18] Birds, too, are known to engage in cooperative behavior separate from their cooperation in rearing their young, which is often striking.[19] And spiders solve complex problems of maze-escape, which nonhuman primates can't solve, nor can all humans.[20]

ORIGINAL NON-SIN

The extent of the similarities leads to the question: were there events in the past that contributed to these similarities?

Consider the following possibility. Reproduction exceeded replacement levels. Groups grew, split, reconnected, and new groups emerged.[21] There was gradually increasing dispersion among our ancestors. Separate groups are likely to have formed and subsequently evolved independently in response to local challenges of economic and military survival on one hand and reproductive prosperity on the other. This scenario would account for some of the differences we have identified. But it's also likely that there were many similar challenges such as personal survival, successful reproduction and rearing of offspring, protection from predators, assuring adequate food, maintaining health, and, increasingly, managing social competition and conflict. These are likely to have led to similar, if not at times duplicate, evolutionary solutions.[22] These commonalities alone could explain many of the similarities in behavior and genetic profiles, and not in a very complicated way, either.

Then what can be made of these similarities—the common physical attributes and ways of organizing and managing lives, behavior, and emotions? From afar they direct attention to the possibility of the shared scaffolding that we have repeatedly emphasized. Moreover, the findings can be organized in the context of the signature behaviors of most religions: socialization, rituals, rules of behavior, hierarchies, and deference to a higher authority or idea. Chimpanzees groom and make sounds; humans talk and hug. Chimpanzees defer to authority and don't injure infants in their group—they, too, follow rules.

Humans do the same. Chimpanzees engage in rituals surrounding greetings (we do handshakes, though chimpanzee males stroke and evaluate each other's scrotums), authority (salutes), grooming ("what a nice shirt"), and nest making (housewarming time). Humans pray, recite mantras, clean house, clean themselves, and help their neighbors. Both organize their behavior in response to local authorities within their species. And in a mighty roundelay, chimpanzees defer to human authority unless they can avoid it, while most humans defer to a god or an equivalent and we overwhelmingly appear to prefer to do so.

Still, what of the differences that are so colorfully and clearly present? What are their implications? What about the

uncertainties? What about the differences between the astonishing Easter Vatican and a shady midmorning forest clearing
in Africa? Are they sufficient together to cause us to reject the
idea of scaffolding?

Read on.

DNA SIMILARITIES BETWEEN CHIMPANZEES AND HUMANS

Even for the most entrenched and vocal skeptics of the proposal that humans and chimpanzees shared a common
ancestor, the first reports of comparisons between the DNA of
humans and chimpanzees were eye openers. Ninety-eight percent the same. There was something there the skeptics had not
and could not have seen.

The skeptics' doubts were not unreasonable. A separation
of six to eleven million years obviously favors the prediction
that there will be genetic differences—perhaps significant differences—despite a possible common ancestor. After all,
today's humans have undergone significant changes from their
ancestors over the same time period. And the skeptics were
right. There are such differences.

The skeptics were also right in predicting that the final act
of the DNA drama was still to be written. For example, there
are still evolving views about genes and DNA—what they are,
what they do, and how best to define them. Technically, a gene
is a chromosomal segment responsible for making a functional
product, which is usually a protein. Or, more generally, the
genome is the DNA and the expressed genome is the RNA
that's the basis for proteins.

What is only now becoming clear is that genomes and
genomic change are far more complex than originally envi-
sioned.[23] As definitions and analytic techniques have been

refined, the original 98 percent has dropped to somewhere between 94 to 96 percent, a difference that represents approximately fifteen million changes in the genome since the time today's two species were one.[24] In part, the percentage changes are due to RNA, the critical importance of which was not grasped in the early moments of DNA discovery. Further, gene location, gene production rate, differences in genetic structure, and the consequences of each are issues still under intense investigation. These studies suggest that part of the human-chimpanzee genetic differences has to do with those parts of the genome that don't control which protein is produced but rather where it is produced, how, in what quantity, and especially, why. And, as with humans, these processes are not exact mirror images of each other among any two chimpanzees, which accounts for much of the individuality among chimpanzees.

Still, the current estimate of 94 to 96 percent duplication is more suggestive than not. It can't be overlooked or dismissed easily.

To make matters yet more complex, another page in the DNA story is just now unfolding. Not only does it appear that the DNA in humans has changed significantly over the past fifteen thousand to five thousand years but also that there have been far more changes among those genes responsible for the human brain compared to those for the chimpanzee brain.[25] Whew! "Extraordinary fast evolution," as some have described these events, appears to have set a world record of expansion, perhaps stimulated by efforts to outsmart both competitors within our species and the many predators without.

How do these points about DNA add up? One possibility is that both past and present genetic changes have had far less influence on the behavior of chimpanzees and humans than on their brains. If so, this could mean that the scaffolding idea is applicable primarily to shared behaviors, less so to their brains. But other findings strongly suggest the opposite.

BRAININESS

We also have to consider brain anatomy, brain function, and brain chemistry. These critically influence the organ's capability and operation. We know two facts of anatomy: the brain of an adult chimpanzee is about one-quarter the size of a brain of an adult human. Some functional centers have different locations. Unfortunately, this is about the sum of the reliable findings. Lack of reliability is in part due to changes in the scientific questions that attract investigators. These are largely responsible for the kind of research that gets done. And research often veers off in this or that direction.

For example, recent advances in skull analysis are forcing a rewrite of human evolution to take account of the many past evolutionary blind alleys and descendants who failed to survive, such as the Neanderthals. Comparison of brain organization and individual brain cells rather than the whole brain or its functional centers also is receiving increasing attention.[26] These studies point to clear differences in the way the brains of chimpanzees and humans are organized, the number of their cell types, and the anatomy of cells.

Then we have to add to these points the possibility of inbreeding during the process of splitting from a common ancestor or ancestors. Such findings inevitably lead to speculations. For example, there may have been several common ancestor splits, not just one, punctuated by all kinds of species mixes, adaptive failures, and a few adaptive successes.

Studies of brain function are in a similar state of flux. For example, there is the hypothesis that humans and chimpanzees think as they do because the brain is modular, the initial structure of the brain is a product of DNA, and DNA has been shaped by evolution. The logic makes sense and particularly so if the species evolved in close parallel. But finding supporting evidence is another matter: except for language in humans, there is no hard evidence that modules exist in the brain, beguiling though the idea has been (although the brain has functionally specialized areas).

Granting these many uncertainties, there are still reasons to suspect a high degree of brain similarity between the two species. For example, human language systems seem to rely on content-addressable memory. This form of memory is widespread among vertebrates. It retrieves information for a variety of functions other than language on the basis of memory content, not its location in the brain. Findings are consistent with this possibility. For example, chimpanzees outperform humans on certain skills. Some have better short-term memory than humans. They use from two hundred to six hundred plants, which they need to remember as well as their seasonality while, on average, humans can identify and will use far fewer.[27] But chimpanzees also appear to lose learned skills faster than humans.

And what of those big-word functional areas of the brain such as amygdala, prefrontal lobes, and hippocampus, which will be discussed in coming chapters? How similar are they in chimpanzees and humans? Do they do similar things? Such comparisons are in their infancy due largely to the fact that chimpanzees introduce simple but confounding technical difficulties not present among humans. Most studies of brain function use functional magnetic resonance imaging (fMRI) techniques to assess which brain centers are active when a research subject carries out an assigned task. The technique requires that subjects hold their head still during such studies, which also involve loud banging noises. Very intelligently, chimpanzees object to this irksome rigmarole and thus defeat even the wisest researcher.

The story changes very little when the focus turns to the chemical makeup and function of the chimpanzee brain. For example, the brains of all primates appear to be far more chaotic than previously imagined.[28] At one time, it was thought that neurons release their signature chemicals only at their ends and that the chemicals triggered the firing of the next neuron in a chain—in effect, a very orderly and highly directional process. But it turns out that some neurons release their chemicals all along their axis—much like a soaker hose in a well-planned garden. This activates adjacent neurons that fire and decrease order and introduce multidirectionality. The full implications of these findings remain to be worked out. But no one suspects that they are going to make understanding the brain simpler.

Some things are known, however. For example, humans appear to have less genetic variation than chimpanzees for a specific type of dopamine—D2—receptor.[29] On the other hand, they appear to have duplicate genetic structures for the dopamine D4 receptor.[30] Studies show that humans and chimpanzees possess similar receptor-binding properties for the brain chemical serotonin.[31] These are particularly interesting findings not only because dopamine and serotonin and their interactions with various features of religion take center stage in coming chapters, but also because far more is known about their chemistry and function among humans compared to chimpanzees.

HOW CHIMPANZEES CHILL

How do chimpanzees try to avoid and reduce stressful and aversive experiences? Here the answer seems clear: they do many of the same things that humans do.

There are three critical messages associated with this. The act of living is stressful. The human brain imagines possibilities and asks questions it can't confirm. This too is stressful. The brain likes answers. And people have developed ways of offsetting and avoiding stress. Are there chimpanzee counterparts?

Chimpanzee life is stressful, at times exceedingly so. Nowhere is this more apparent than in the dynamics of a group's hierarchy. For example, to be a dominant male is to live with constant challenges from subordinate males that would assume the dominant position if only they could. They are always running for election. To be subordinate means living by irritating rules such as: good things to eat first go to the dominant animal should he want them, females will be less receptive to subordinate males' overtures for copulation than to those of the dominant male, and attempts at copulation by subordinate males may invite the dominant male's wrath.

It is broadly necessary to act submissively in the presence of a dominant animal and his allies. Otherwise there could be physical and psychological consequences. Still, on balance, it seems true that as stressful as group living is, it's apparently preferable to solitary existence. As with humans, there are few chimpanzee loners. There is also the potential threat of other marauding groups and an occasional large and hungry member of the cat family. Chimpanzee groups raid other groups, often killing their males and offspring, and capturing their females. In effect, they terminate the life of the group they have so enthusiastically attacked.

Group living requires ways of reducing stress. Postconflict resolution has been mentioned—animals that were enemies moments before seemingly forget their differences and engage in reciprocal schmoozing. Friends position themselves close to friends. Groups made up of members that may be competitive at other times of the day often forage together cooperatively for food or assist each other to prepare sleeping sites.[32]

It almost seems as if the group is enjoying something like a religious holiday. Actions are slow and nonthreatening. Vocalizations are subdued and modest and conviviality the accepted norm. The animals have chosen a protected and calm forest location. The air appears to have a special lucidity, rather like the ambience of great cathedrals whose size appears to expand the power of light and silence. Among the chimpanzees the natural space appears to provide them shelter, though of course nothing has been built. The ongoing cares of daily life are somehow muffled. When animals are relating in this way or sleeping, challenges, threats, chases, and fights are left to other moments.

SCAFFOLDING

This brings us to a closer look at scaffolding, which is in principle a very simple metaphor. Looked at from afar, it's analogous to two buildings. Each has a similar internal structure—I-beams, pipes, electrical wiring, and so on. But there are outward differences in detail. One has shaded windows while the other has clear glass windows. A closer look reveals further differences. One building has ten elevators and a cafeteria on the third floor while the other has eight elevators and a coffee shop on the first floor. Still, both have windows, elevators, and places to eat. The functions inside also will likely differ somewhat or a great deal. One building houses a stock brokerage, the other a chocolate truffle factory. Yet greater detail doesn't necessarily assure further differences.

For example, we now know that Old World and New World nonhuman primates and humans have very similar complex hardwired behaviors. These include aggressive facial patterns, defensive forelimb movements, hand-to-mouth coordination, and reaching and grasping. These are highly similar across species, which suggests there has been minimal genetic change between humans and chimpanzees over time.

We also now know that the way proteins are developed from DNA is often strikingly similar across the two species and, most dramatically, highly similar in the brain. Further, comparative studies increasingly suggest that chimpanzees engage in a high-level decision-making process in which they employ criteria requiring controlled cognitive processes.[33]

So where is the bottom line in this discussion?

In our view, the similarities between chimpanzees and humans in their appearance, anatomy, behavior, DNA, and in their stress-relief strategies make it highly probable that they share a brain-body-behavior scaffolding. This scaffolding serves as the basis for religionlike behavior among humans. Put into another perspective, with the exception of the elaborated belief features of religions (e.g., ideas contained in the Bible, notions of pilgrimage), the fundamental behavior characteristic of literally all religions—socialization, rituals, rules, deference to authority, hierarchical systems—are more similar than not among chimpanzees and humans. Among both species, the scaffolding directs and constrains the types of behaviors that occur and which account both for their similarities and differences. Detailed findings addressing brain anatomy, chemistry, and function remain to be gathered to fill out the picture.

Also in our view, it's onto this scaffolding that humans have built, elaborated, and added products of their imagination and often complex religious beliefs about things they don't experience directly, such as an afterlife and an authority higher than the human authorities they encounter each day.

Without primatological scaffolding, religion becomes even more improbable than at first glance.

Now to a more vexing question: do chimpanzees have imaginings and beliefs? Findings from the study of language are not encouraging if one is hoping for a "yes." Chimpanzees living in laboratories under the guidance of humans can learn up to two hundred words, solve simple word problems, invent novel word combinations, and issue and follow word commands. Getting them this far takes very considerable effort. However, they don't develop these capabilities on their own in the wild. So language, which is an indispensable part of human religion, is largely absent. Still, there is reason to assume that they imagine and believe. They do have culture. They modify their behavior according to group norms. Moreover, there are a host of findings pointing to the conclusion that nonhuman primates have brain processes that are integral to imagining and believing.

Studies suggest that humans think and develop ideas before they learn to talk, thought precedes the acquisition of language.[34] Similar findings apply to nonhuman primates. For example, various studies have shown that neurons in the parietal cortex represent and process the relative value of competing actions before they are taken.[35] Dorsal premotor cortex neurons rehearse what an animal will do before carrying out a well-learned action.[36] Observing others in action has its representation in the frontal lobe even though no action may follow. So it's very possible that chimpanzees think and have beliefs even though they are not good talkers and have not gone in for formal theology.

Which brings us to morality.

MORALIGION

Morality and religion can be separated by definition. However, in practice there is significant overlap. If it is about nothing else, religion is largely about good and bad behavior, both in private and in the social arena.

Christopher Boehm of the Goodall Center and the University of Southern California captures many of the points for chimpanzees:

> What should be indisputable . . . is that captive chimpanzees can understand the difference between conduct that is

approved or disapproved in their human master's eyes. So, in their own way they understand our rules—which is a basic and very important part of our own moral capacity. In the wild they learn rules in that they come to know which of their behaviors will predictably get them in trouble with a dominant. They also learn to greet a dominant individual with a submissive pant-grunt, and to defer if a prime feeding place is available and there is only enough fruit for one.[37]

Whatever else these observations mean, they suggest that chimpanzees can learn and behave through understanding and incorporating a system of rules other than those they develop on their own, as if they have been taught, say, Calvinism. And it's clear they have beliefs as, for example, the belief that behavior X will result in consequence Y.

The idea of chimpanzee morality is consistent with many of the ideas expressed by scientists. For example, there is a growing consensus among evolutionary biologists that there is a biological underpinning to morality and moral practices among humans. That is, human morality has primate origins. For example, Franz de Waal of Emory University, in discussing the animal roots of human morality, argues that animals have an inbuilt sense of fair play and can show compassion to others in their group. They exhibit what he calls precursors for moral behavior. From such observations it is not a long jump to the possibility that chimpanzees have their own form of religion, minus perhaps the idea of a higher authority. Recall that Buddhism has no god, no explicitly imagined higher authority.

There are, of course, many types of morality. Three are especially important here: the effects of one's behavior on others; what one does alone, which others can judge; and what one does alone, which others don't know about. Religions generally emphasize all three types—there is no escape behind the bedroom door. As in a French farce, closing that door invites opening the one to the confessional booth.

And both humans and nonhuman primates often play one type against the other. For example, in laboratory experiments where monkeys can observe each other and perform the same task but receive different rewards for success, monkeys that receive a lesser reward may refuse to play. Among humans, there are saints in public who are sinners in private. So too with nonhuman primates, which, for example, will steal from other members of their group when others are not watching. Again, the similarities are striking. That much of this behavior is brain-based is suggested by findings showing that humans suffering from damage to the ventral medial prefrontal cortex are sharply transformed from their preinjury moral judgments. Part of the morality machinery is broken.[38]

Then there are the ideas of Robin Dunbar, who has argued that religion wasn't there in the beginning and that humans invented it primarily because it facilitates group cohesion.[39] Invented, it differs from David S. Wilson's idea that a religious trait evolved as a group selection outcome.[40]

In essence, the question is: does a religious trait improve the evolutionary success of individuals who sustain it, or rather, the evolutionary prosperity of the group of which the individual is a member? Overall, while there is clear association between the two interpretations, the first one seems more durable. The scaffolding idea is closer to Dunbar's view than to D. S. Wilson's.[41]

But there is a critical difference. The essence of what we are suggesting is that chimpanzees and humans share a highly similar brain-body scaffolding. Both species have evolved to imagine and believe, though human evolution has noticeably outpaced that of chimpanzees. Humans have evolved so that they can articulate what they imagine and believe in ways chimpanzees can't. Their imaginings and beliefs take into account a vastly greater knowledge of the real world and a variety of personal experience than those of chimpanzees.

They are also far more advanced in creating imaginary worlds that are unknowable.

Some, but far from all, of the imaginings and beliefs that humans have are attractive to others. They may readily facilitate group cohesion. But some may do the opposite. Some of these may become the bases of religions. When this happens, the forms that religions take are directed and constrained by their behavioral scaffolding.

Chapter 7

MY BRAIN. YOUR LITURGY.
OUR STATE OF GRACE.

S tress is real. It is an unavoidable product of living in a real
world and with other beings. Irritation, difficulties concentrating and recalling, confusion, worry, anxiety, listlessness,
fatigue, a poor balance between moments of pleasure and
dismay, hypersensitivity, sleeplessness, a negative outlook—
these are a few of its signs and symptoms.

They are familiar to everyone. No one is immune.

Many sources of stress are banal. However, this doesn't
mitigate their effect. Many derive from the items that fill the
menu of everyday life. There is stress that accompanies uncertainty and doubt about the future such as paying off a debt
and the possibility of an afterlife. There are workings of the
brain over which owners have no control but contribute to
stress nonetheless. There are stress-initiated chemical and
functional changes in the brain and the body that are responsible for the signs, symptoms, and the somatic consequences of
stress, even at the level of the gene.[1] There are strategies people
employ to avoid and minimize stress—take a course, do group
therapy, watch a ball game, hike. And there is religion.

This chapter rummages in the church closet, the door to which many prefer closed.

FAMILIAR AND UNAVOIDABLE SOURCES OF STRESS ARE A PART OF EVERYDAY LIVING

Perhaps surprisingly, the most frequent causes of stress are seldom due to life's milestones, such as the birth of a child, graduation from college, marriage, and one's fiftieth birthday. Usually these events occur years apart. They are anticipated. There is advance planning and preparation. There is helpful lore about them. Stress is part of their design, and a sense of accomplishment when they're done is a durable reward.

The most frequent stresses have their origins in everyday goals and the efforts to accomplish them. People organize their lives and their behavior to achieve specific objectives, such as shopping for the family, plowing a field, meeting a work deadline, cleaning the garage, minimizing the uncertainties that go with living, and avoiding serious accidents and diseases. Circumstances that interfere with these objectives are unwanted. They irritate and frustrate, and they trigger changes in the brain and the body. At a certain point, irritation becomes a medical issue.

Take a day in the life of Mr. Powwow. He rises, prepares and drinks his usual cup of coffee, eats a duet of hard-boiled eggs or somewhat inventively scrambles them with cheese and green peppers. He considers taking a shower, dresses, and departs for work in his car. En route, a delay due to an automobile accident assures that he will arrive at the office late. At a midmorning office meeting his proposal to change endlessly vexatious office procedures is rejected and, even worse, an incomprehensibly bureaucratic rigmarole is strengthened. At noon, he receives a telephone call from the nurse at the local high school telling him that his son had been hit by a baseball bat and is in an ambulance headed for the local hospital. After visiting his son, it is too late to return to work. Once home, he sits down to rest and read the day's mail, in which he finds a letter from the town lawyer informing him that he is being fined a thousand dollars for cutting down a dead, rickety tree on his property without the city's permission.

Or take a day in the life of Ms. Zanzibark. She rises, pours herself a cup of Scottish breakfast tea, and prepares cereal, juice, and toast for her daughter's breakfast. She changes from her bathrobe to her school clothes and gets in her car to travel to a nearby school, where she teaches the third grade. Her morning proceeds quietly. Students are cooperative and have mastered their lessons. At noon, she and a friend drive to a nearby restaurant for lunch. As they depart, she notes a size-able and new dent in the left rear fender of her car. Back at school, there is a message to call the principal. They meet. A discussion follows. It's about a complaint by a fundamentalist employee who has accused her of making derogatory remarks about her religion. She was overheard saying, "There is very little hard evidence supporting religious beliefs." A formal hearing is scheduled for the following week. She has the option of bringing her own lawyer to the meeting. More cost, in addition to paying to repair the dent in her fender.

The reality of everyday life is that not all days are like those experienced by Mr. Powwow and Ms. Zanzibark. Some are less stressful. Some are more. From birth to death, no two days are quite the same. It could be the weather that interrupts plans, faulty machines, unexpected road work on a vital highway, guests, next-door neighbors, illness, children, spouses, relatives, credit card companies, low birth rates among polar bears, a national emergency. The list is long and endless, as extensive as the calendar. Supposedly good news can turn things bad, as many a lottery winner knows. And at times these unexpected and unpredicted events arrive with high and unpredictable velocity. Up to a point, the brain and the body are resilient and take them in stride. But the brain and the body can bend only so far before their real and metaphorical fibers begin to break apart.

What is predictable about today, tomorrow, and the next day is that they rarely unfold as planned. One gets up in the morning and fuzzily or precisely charts the day's goals, or maybe it was done the night before: first do A, then B, then C, then D, E and F, and so on. Hopefully each step will unfold like clockwork—who regularly plans for a flat tire, an unexpected lawsuit, a robbery? Moreover, goals change, skills are refined, motivations vary, and the environment and the options it offers are always in flux. What was a goal last year is not necessarily a goal this week. What was possible last year may be easier or more difficult to achieve today. *You can't twice stick the same foot in the same river* is both an Indian and a Greek proverb that captures the details of daily plans and their unfolding.

Isn't the cure here contingency plans? Some devise such plans to counter the possibility that the day's plans may go awry—extra money in the wallet, spare tire checked the last time one filled the car with gas, and understandings among close friends who will help in emergencies. Wise planners, no? Yes, of course! But these plans, too, can go awry. Add the fact that there are significant individual differences in how unwanted events and circumstances are handled. Sex, emotional intelligence (one's skill at interacting), personality, skill differences, and more each contribute. A consequence of these differences is that people differ in their ability to navigate the social environment; to integrate and manage their thoughts, emotions, and motivations; and in their vulnerability to events that alter their brains and bodies. They generate differences in stress.

There is simply no certain and obvious escape, no certain

formula or strategy to avoid stress. Color, creed, nationality, bank account balance, prior experience, or social status—no one is immune, though low social status appears to heighten feelings of stress.

Living has its unavoidable costs and abrasions. Living in the valley between fond reverie and oncoming reality is an inevitable foundry of stress.

This is about reality, which is assertive enough. Then there are imaginings and beliefs and doubts about unknowns. They, too, are a source of stress. Some will resolve themselves with time. For example, newlyweds wonder whether they will have a happy marriage. A minister assigned to a new parish wonders whether he will be accepted by its members. The investor who shifts his assets from equities to real estate wonders if the housing market will crash. And parents wonder whether they will raise their children successfully. Time usually tells, or mumbles something.

Such uncertainties are not the primary topics here, however. Our focus is on uncertainties and unknowables that fall under the umbrella of religion, all those quite remarkable constructs such as, Who am I? What is the meaning of life? What happens when I die? Is there a soul? Is there a heaven? Was Christ really resurrected? Is there a hell? What is God really like? Does He have a personality? How were people created? Does the way I act while living affect my options or lack of them after death? Will there be benefits here on earth if I act according to God's wishes? Such questions perhaps hint at why many church services have a large number of grey-haired people in attendance.

Each and all are questions about what is unknown. Questions that can't be answered with certainty except by personal conviction. None are supported by hard evidence. Yet none will soothe the brain or lessen stress if left unanswered. Pascal is said to have summed it up this way:

I feel engulfed in the infinite immensity of spaces whereof I know nothing and which know nothing of me. I am terrified. . . . The external silence of these infinite spaces alarms me.

We have all asked such questions. Usually, infrequently early in life—it's the privilege of youth to be committed to other matters—and more frequently with age. They reflect both an intense curiosity and an ambient sense of shimmering imprecise danger about life and beyond. It's the ambient sense of imprecise danger that is chronically stressful.

And where do these questions come from?

Earlier we mentioned possible sources, ranging from the naturalistic idea that the brain has evolved such that it naturally generates such questions to the theological assertion that a god instilled such questions and their answers in the brains of humans. Or they may have their origin in everyday experience. For example, children love their dogs and cats and usually outlive them. And, unlike lizards or worms or flies, their dogs and cats have names, personalities, habits, and preferred sleeping places. Owners often wonder what happens after they have put their companions to rest. There are dog and cat cemeteries but none for worms or lizards or flies. Years following their deaths and burials, pets are remembered, their antics discussed, and, most important, they are missed. In effect, they enjoy a kind of afterlife. Might there be a human equivalent?

There also is upbringing. Parents and others teach children—indoctrinate in some instances—that there is a specific god, or an irreplaceable spirit, or a totem that requires respect and dictates that certain behaviors are essential to assure one's chances for a satisfying life and afterlife. How many grandmothers have taken the admonitory opportunity to advise their grandchildren that brushing their teeth, doing household chores, and behaving as their mothers and fathers expect are essential for a ticket to heaven?

Nor are such teachings without behavioral consequences. For example, there is experimental evidence regarding what has been referred to as the *Macbeth effect*.[2] This is a threat to one's sense of moral purity or an act that compromises the sense of one's moral integrity and induces the need to cleanse oneself physically—literally, wash one's hands more often— and mend oneself ideologically and psychologically. When the concerns and fears about unknowns are combined with the teachings of youth, a person becomes the captive of beliefs and doubts from which there is no easy exit. Witness the anguish that often accompanies a person's rejection of his religious upbringing. This experience has been no more fully illuminated, as we have already noted, than in James Joyce's *Portrait of the Artist as a Young Man*, which describes the transition from believer to the artist who defines his own sentient and moral world.[3] Or as William Blake once announced, "I must create a system or be enslaved by another man's."[4]

Why would so many people preoccupy much of their lives with such questions and doubts? Why might people believe the answers they are taught about unknowns? After all, the average person discards much of what his parents and teachers taught him by the time he is twenty years old and goes about forging his own beliefs, biscuit preferences, values, and life plans.

An answer to these questions is found in what the brain prefers—in what is tasty and tangy to the brain. Religion pleases the brain's sweet tooth. Besides some short-range exceptions—most people don't want to know the outcome of a sporting event before it is over, otherwise, why attend?—the brain strongly prefers certainty over uncertainty, resolution over open-endedness, and balance and symmetry over imbalance and asymmetry. Cups and saucers are stacked in order in the cupboard. Tools have their special place in the work room. Plants in the garden are not just placed randomly. Budgets are balanced. Pyramids are symmetrical, as are places of worship. There are tacit lists of good and bad friends.

Even questions and doubts demand a symmetry. Questions should have answers. Doubts should have resolutions. Stories need endings. Order, balance, and symmetry make the brain comfortable. It's these outcomes toward which humans strive. And for most people, providing an end to a story, resolving a doubt, and giving some order to uncertainties is far preferable to wondering and hoping but not knowing for sure. Unanswered questions, unresolved doubts, too many uncertainties —all generate stress, at times literally unrelenting and debilitating stress.

There is a related point here. It is this. Everyone believes a bit differently, particularly in how one resolves uncertainties and doubts. The point applies to members of the cloth as well. Recall their often bitter and decades-long doctrinal disputes. But what might such differences mean? Is there anything more to them than saying that because people differ, so too will what they think, feel, and believe? The answer is yes, individual differences count. For example, the person who believes that committing a sin may mean he or she will be denied admission to heaven and the person who views the same sin as a guideline for relatively casual daily behavior will have very different stress responses should they sin. In the former, worry, guilt, and preoccupation with one's personal and religious credentials and one's future are likely to continue long after. In the latter, perhaps at most, the sin serves as a reminder not to repeat it.

BRAINSOOTHE AND OUR SOCIAL BRAINS

If the events of day-to-day living and our imaginings and beliefs about unknowns are not enough to ensure stress, the brain has its own characteristic operations that frequently add to stress. For example, there is a part of the brain called the amygdala, a brain center that processes emotion-laden messages from others and that initiates emotional states and responses in the recipient.[5] It is particularly responsive to the direction of others' gazes when they send messages of anger or fear.[6] Oddly, messages sent head-on cause significantly less amygdala activation and emotional response than the same messages sent when the sender's head is at an angle. The degree of exposed eye-white—squinting versus eyes open— leads to similar differences in amygdala response.[7] The wider the eyes are open, the less activated the amygdala and the less the recipient's emotional response. The reason for this remains a mystery. The old military adage comes to mind: Don't shoot until you see the whites of their eyes.

Other studies show that when an observer views a person suffering from pain, his brain responds by activating the same brain centers that are activated when the observer himself experiences physical pain (anterior cingulate cortex, insula, cerebellum).[8] It's a conversation between brains.[9] Much the same is true for rejections by others.[10] Experiencing rejection initiates activity in the same functional centers as those activated when the rejected person experiences physical pain. This is how and why films and plays achieve their impact. In effect, the brain has its own inbuilt ways of processing others' communications. These sharply affect how its owner perceives, feels, thinks, and responds. And, much as one might try, the brain resists training to do otherwise.

The urgent and central point being made here is this: the human brain is social. It is not about pure, independent thought. It is never fully free of the influences of others on the way it functions. And, clearly, one can't fully control how others behave and communicate. Having a social brain means that the usual view of the brain—mine in my private head, yours in yours—is only correct anatomically. Its operations pay little attention to its physical boundaries. Try ignoring a spouse's request, a nasty remark from a colleague, an admonishment from a minister, even a momentary encounter with an unknown beggar asking for only a few pence just after you've deposited a sizable check in the bank. Of course, others very much affect how we think and feel.

It's these influences by others, and especially aversive influences, that people try to control largely through carefully selecting their social colleagues and their social encounters. And there are good reasons why. As we will discuss in detail in later chapters, positive, complimentary, and affirming messages from others not only are pleasurable, but they also have salutary and stress-reducing effects on the brain. They *brain-soothe*. Such messages nudge the brain toward its optimal chemical and functional states. The opposite is true for negative messages, where one is ignored when seeking the attention of others, a kind of negative inaction by those who fail to respond. These points readily relate to religion and its practices. Exchanges between rabbis, priests, ministers, imams, and members of their flocks are exchanges carried on by a social brain. When they are positive, that's fine. When they are negative or demanding, it's another matter. Stress is one outcome.

But the story of others' behavior and its influences is still far from complete. Reciprocity, deception, hierarchy, and belief also are involved.

As we have noted, social relationships among unrelated individuals build on the exchange of favors and helping—reciprocity.[11] Generally, the closer and more lasting the relationship, the more frequent and reliable are the reciprocal exchanges between the participants. Fair reciprocators are predictable helpers and are trusted. Unfair reciprocators are unpredictable helpers and they are not trusted. Where possible, unfair reciprocators are avoided. In extreme cases, they are socially ostracized. In effect, play by the rules of social exchange or you are not a friend.

Reciprocity, or at least its potential, is not limited to exchanges between two individuals. People also observe the behavior of others. Are they reciprocators or not? Answers one way or another affect the brain. For example, empathetic neural responses are modulated by the perceived fairness of others.[12] The social brain is on the job again. Judgments follow, which influence the reputations of those observed and their possibility of gaining new friends.

There is also deception. People send deceptive messages about their intentions and knowledge. Deception is only possible because most messages that are not deceptive are generally accurately perceived. It's this fact that makes occasional well-crafted deceptions more likely to be effective. And often they are. We have all been victims. At times, even seemingly close friends betray one another. Where would countless playwrights and novelists be without that reliable plot line? Nonetheless, over time and with experience, most people develop a degree of skepticism about others' stated intentions and their behaviors. People thus are often alerted to the possibility that others may say one thing but mean another or that a seemingly helpful behavior is undertaken with an ulterior motive. Still, no one bats a thousand in defending against that cheating heart.

Religion relates to these points in several ways. The first deals with reciprocity and hierarchy. Relationships between religious authorities and members of their flock are a form of reciprocity. Consider confession among the Catholics. Those who confess reveal intimate details of their lives, details often hidden from close family members and friends. In turn, priests provide discretion, guidance, forgiveness, and reassurance to those who confess. Similar reciprocations are repeated daily in religions the world over. Informed forgiveness is the coin of the realm.

THE GREAT CHAIN OF BELIEF

The second way deals with belief, deception, and hierarchy. Although atheists and members of different religions may be highly skeptical, even openly critical, of the practices and beliefs of other religions and their members—how many Christians believe that a group of virgins awaits them as a reward for religious martyrdom as some Muslims do?—members of religions generally do the opposite. They grant unusual authority to those high in their religion's hierarchy and they readily accept their religion's prevailing dogma. They believe, take advice from, and generally revere or at least respect those members of their religion who have been ordained with high status. Such individuals exhibit a surprising willingness to relinquish their normal skepticism about the possibility of deception. Surely some internal mechanism makes this possible.

And why might believers be so willing to grant authority and insight to religious authorities? A key point here is that both deception and its detection are strongly influenced by social rank. The higher a person's social rank, the more likely his deceptions will go undetected (apart, of course, from such formal detectives as members of the press and more recently tax authorities). Conversely, the lower a person's social rank, the more likely he will not detect a deception and the less likely his deceptions will be successful. An obvious implication of these points is that those high in religious authority are well positioned to deceive, especially if they deliver the news that their messages derive from a supreme being to which everyone is equally beholden. Conversely, those lower in the hierarchy are positioned to believe. A factor in their willingness to do so is their suspension of the precautions against deception that are available to use. Again, the social brain at work.[13]

There is still more. Reciprocity ascends to another level of complexity when one attributes a personality to a god or some other imagined higher authority. This is especially so if one believes that certain behaviors are required to please one's deity. Parameters change dramatically: there is a unique form of reciprocity between a member of a religion and his god. By working, fighting sin, proselytizing, living according to a religion's rules, and so on, a member helps his god. In turn, his god provides special favors, both in the present through one's certification as a member of his flock and possibly, even presumably, also in an afterlife. Most important perhaps, concerns about deception disappear. Would a god deceive?

And where are the brain and stress in all of this?

STRESS TASTES SOUR
TO THE BRAIN AND THE BODY

How does the brain function when it is stressed?

We will identify some fundamentals that outline the framework and then the details within it.

First, the brain absorbs information from the physical environment, from others, from the body, and from itself. This is, after all, what memory is about. Memory is the brain's pantry. And it does these things largely without a person's awareness. Recall the influence of another's gaze on the activity of the amygdala; the perceiver is conscious only of the angle of another's gaze and his own response, not the specific events occurring within the brain. But the significance of the angle of gaze of others is only one of literally hundreds of examples of how our brains work below the level of consciousness. (It is perhaps a gift of evolution that the brain has turned out this way. Imagine what days would be like if we had to consciously initiate, experience, and analyze each brain action. Compare this to asking a centipede how it walks.)

Second, it processes the information it receives. It organizes, orders, categorizes, prioritizes, connects, and explains, also differently among males and females.[14] For example, a sudden sharp pain in the leg might suggest a bee sting if one is working in the garden but a cramp if one is lying in bed indoors. In effect, the brain creates new information on its own. Emotions, body feelings, thoughts, explanations, imaginings, and beliefs are the outputs of this process.[15]

Third, it makes decisions and initiates actions, such as developing the plans for the day, as we discussed earlier. And plans usually emerge. For example, it's best to mow the lawn before raking it, to decide on the guest list before Friday's dinner is set, to plan a vacation before leaving home, and so forth. Actions follow.

Fourth, throughout all this, the brain assesses the effectiveness of its decisions and actions. Plans for action are accompanied by expectations of their outcome. People don't act randomly without any expectations of what might happen. An effective decision-action is accomplished when the outcome and the expectation closely approximate each other. An ineffective decision-action reveals a discrepancy between expectation and outcome. Effective outcomes such as achieving desired goals are usually minimally stressful. Ineffective outcomes range from moderately to intensely stressful.

Of course, far more takes place in the brain than is being described here. Moreover, the terrain is littered with paradoxes. For example, the worthy goal of visiting a dying friend is likely to be extremely stressful. Much the same is true for complex tasks such as reorganizing a company or even a thorough spring cleaning of one's home. And the frequency of these types of situations should not be underestimated. We all live in a web of social relationships with a mixed bag of obligations. When we carry them out even while we are motivated to do so, they may add to other sources of stress. For many of these moments the brain is accommodating. But there are limits.

The brain does much of its work chemically and it's a real phenomenon. Small molecules jump from one nerve cell to the next to transmit information or to influence its transmission. Depending on what the goal of the brain is, these chemical events occur in different functional centers of the brain that perform different information processing and action-related tasks. To complicate matters, many of these functional areas are networked to each other. For example, just behind the forehead at the front part of the brain are the frontal lobes, which are largely responsible for thinking and decision making.[16] Further back in the brain is the hippocampus, which is a critical system for memory storage and retrieval. The amygdala has already been mentioned; its pivotal function is to interpret emotional messages and initiate emotional responses.

Though crude, we may suggest that what happens in the brain is much like what goes on in the kitchen of a restaurant. There is the refrigerator for storage of things that need to be kept cool. There are the cupboards full of spices and herbs that will influence the taste of what will be cooked and eaten. There is the stove for cooking things. There is a drawer with utensils to assist in the preparation of the meal. And menus change. The meals that result reflect the contributions of these many elements, all different but all essential.

What goes on in the brain is not without cost. There are unavoidable costs in carrying out its tasks even in minimally stressful situations, just as there are unavoidable costs to enjoyable physical work, say, a walk on a mountain covered with wild spring flowers. Small molecules are metabolized and need to be resynthesized. Some have to be transported from their source in synthesis to their location of action, where they do their work. Functional centers can become overworked and suffer from neural fatigue—try memorizing a poem for an hour and note the decline in efficiency after about twenty minutes. Much like tired muscles or when one experiences days like those of Mr. Powwow and Ms. Zanzibark, the brain requires time and chemical change to recover from its work and to offset the effects of stress. Again, this is a very real phenomenon. The brain is energy greedy.

When the brain is stressed, a series of chemical changes and functional activities take place in both the body and the brain. For example, there is an increase in the body's hormone adrenaline in response to the acute stress of frightening situations. Simultaneously, brain pathways that suppress pain may be activated.[17] Such responses suggest energy-costly preparation for flight or fight—that is, attack or flee the source of the stress.

Separately, back in the kitchen, there are elevations in the hormone cortisol and the brain chemical corticotrophin-releasing factor, which are laboratory proxies for the level of anxiety.[18] These occur more slowly than the changes associated with acute stress. Elevations take longer to develop and are far slower to resolve. Let's say the stressful conditions continue. Then other chemicals such as dopamine and norepinephrine increase in the medial frontal cortext.[19] This is a functional center also associated with thinking as well as decision making and memory. Each of these changes is a sign that the brain is involved in alterations that compromise its ability to function optimally. And hand in hand with these changes appear the signs and symptoms of stress—worry, difficulty concentrating, sharpness with others, accidents, fatigue, and so on.

This is just a sampling of what goes on in the brain and the body in acute and chronic, or prolonged, stress. Perhaps one hundred different chemical changes are initiated by stress. At least as many different functional centers and connecting pathways are involved as well. Welcome to a bustling kitchen during the lunch-hour rush.

If stress becomes prolonged, other signs, symptoms, and body changes occur, such as serious and often debilitating body sluggishness. There may well be suppression of the immune system, hypertension, decreased levels of critical hormones and chemicals associated with normal functioning, and, of course, elevated levels of stress-related chemicals and hormones. Robert Sapolsky of Stanford University has tracked these both in humans and other primates.[20] In extreme instances of stress, brain degeneration can occur along with cell death. Even voodoo death (when the brain attacks the body) may reflect this process at its extreme.[21]

As with literally all things human, individual differences play a part. And stress is no exception. Some differences are explained by different genetic makeup and their effect on brain chemicals and their targets. The chemical serotonin, which we will say more about in later chapters, is an example. Different genetic endowments directly affect its efficacy. Furthermore, mild levels of stress seem to activate some people but have the opposite effect on others. Social factors are also important. For example, in response to stress, high-status males do not have the sustained levels of cortisol that the majority of males have. Status thus may offer some tangible protection from the aversive effects and chemical and functional changes associated with stress and the illness that may follow. This was dramatically shown by Sir Michael Marmot in his study of the different health experience of high-status civil servants and low-ranking ones in the British government both having precisely the same healthcare from the National Health System. Those of high status fared far better.[22]

High-status baboon males show increases in the sex hormone testosterone in response to stress, but subordinates show declines.[23] Testosterone is known to dampen the sensation of pain in human males. (Suggestive as this association is, however, testosterone elevation in response to stress has not been established among high-status human males.)

There is also the influential fact that people may not be aware of a brain under stress or that personal behavior is contributing to it. For example, take attending a football game. Say the game is close. There is the physical effort associated with rooting. There is the suspense and uncertainty about the outcome. And there is the relief or agony depending on the outcome and one's preference. The event is exciting and absorbing. One may be delighted to have experienced it. But the brain work that is involved can be immense, as often indicated by the fan falling to sleep after the contest is over. He has been stressed. And we know that in males, at least, winners and their supporters secrete testosterone, which feels good, while losers are noticeably depressed physiologically.[24] The sports pages reveal the chemistry of the brain. And they appear daily, so that brain chemistry matters and is in its way its own reward.

Let's sum up. The changes in the brain and the body associated with stress have myriad causes, such as everyday events, doubts and uncertainties about unknowns, and things that the brain does on its own. The brain works—expends energy—when it copes with even the most mildly stressful situations. When an individual fails to achieve goals, meets with personally distressing information, or can't get extricated from complex obligations, the brain often is unable to process the information, manage the situation, or chart an acceptable course without significant costs and alterations. These directly cause it to deviate from its optimal state of functioning.

Enter stress, smack in the center of the kitchen. And very prominently we perceive the brain and body chemical and hormonal changes that go with stress, the signs and symptoms that accompany it, and the progressive decline in efficiency that is one of its consequences.

CAN WE PLEASE BEGIN
THE *BRAINSOOTHING* NOW?

The signs and symptoms of stress are unpleasant. People act to avoid or reduce their effects just as they do with most things unpleasant, whatever their source. They scratch when they itch, eat when they are hungry, and drink when they are thirsty.

If stress is as ubiquitous, pervasive, and aversive as we have claimed, it's reasonable to expect that people have developed numerous ways to avoid and relieve it. And this is the case. There are hundreds of individual solutions, as well as a host of commercial ones. To cite but a few: change jobs, send children to boarding school, walk, hike, fish, jog, vacation, garden, read, sleep, visit with friends, go to the bar or the movies, have an affair, get divorced, shop, play tennis, smoke, take drugs or sip Côtes du Rhône, party, and so on.

The commercial list is equally impressive: massages, spas (the real estate industry is energetically busy producing restful spas), vacation tours, hiring an accountant or secretary, opera, concerts, plays, Broadway musicals, dude ranches, AA, education, gyms, and so forth.

For many, these solutions may be effective. Otherwise they would disappear. Serious joggers would attest to this fact. Massages are at least six-thousand years old and still thrive. Sporting events and entertainment prosper as well. There are interesting and often similar features to many of these activities that help explain why they may offset stress. Many are simple, straightforward activities, as is the case with jogging, hiking, or going to the movies. Often they can be done alone, such as sleeping, or, if not alone, with others whose behavior is predictable and soothing, as with massages and spa experiences or spending a relaxed day at the beach with a loved one. These are activities that other people or unforeseen contingencies are unlikely to disrupt.

Nonetheless, the bottom line is that for the majority of people these solutions are only successful on a short-term basis. One can only jog for so many years, people cancel their gym memberships, and the spa is fine while it lasts. But Ms. Zanzibark still has a principal's hearing scheduled next week. Mr. Powwow's son will be convalescent for several weeks and may not fully recover. No one is immune.

LET THERE BE RELIGION

And now the plot thickens. At this point, religion enters our story in a formidable way. Religious beliefs provide answers, complete stories, and provide order.

Religious experiences and behaviors offset many of the effects of stress. And the conviviality among members and religious authorities soothes the brain. These are the ubiquitous phenomena we will discuss next.

Chapter 8

THE ELEPHANT IN THE
CHAPEL IS IN YOUR SKULL

A versive brain states are those mental sore feet of daily living that unsoothe the brain and the body. They return here to the discussion. *Brainsoothing* returns also with the question: How does the brain soothe itself? That is, how does it jettison the jagged consequences of palpable daily reality? And how does it confront the strange uncertainty that may surround the events following one's medical, legal death?

Socialization, ritual, and belief are signature features of literally all religions. They are the topics of this and the next chapter. If well packaged, this trio of factors provides a kind of neighborhood clubhouse where believers can systematically and with predictability reduce the undesirable effects of stress.

Yet *brainsoothing* doesn't happen without effort. It is certainly not a part-time job.

Somewhere, in many, many somewheres, people are congregating for a religious service, participating in one or on the way home from one. Or they are meditating or praying in private or carrying out deeds in the name of a higher authority.

Or they are wondering whether their behavior is adequate to assure an admission ticket to the infinity they prefer.

In the world at large these events are never ending. They quickly add up. Some rough hypothetical numbers make this clear. If 80 percent of the world's 6.5 billion people who identify themselves as believers spend two hours a day in religion-related activities—a low estimate by many accounts—such as praying, traveling to and attending services, and performing worthy deeds, then *each day* 10.4 billion hours or 61.9 million twenty-four-hour day, seven-day weeks are spent in these activities.

The world, of course, is a relentlessly diverse place and people become believers at different ages. To account for those young people who are as yet uncommitted and those intrepid souls who would chance the possibility that there is no higher authority, the 80 percent estimate might be lowered to 70 percent, or even 65 percent, and two hours per day might be reduced to one hour. Nonetheless, billions of hours per day is clearly a barometer of something very important to the majority of people who walk the earth. It is something they value, something they need, and something they seek. And imagine what the numbers would be if the 80 percent or the 70 percent or even the 65 percent of believers spent four hours per day in religion-related activities, as is common among many believers—for example, Muslims who pray five times daily and often in busy, congested environments, which demands travel time. Organized religion is big business, perhaps the biggest, longest-lasting, and most durable business humans have created—ever.

Now to *brainsoothing*.

The short-but-accurate answer to how religion *brainsoothes* is simple and straightforward.

The brain alters itself through religious socialization, ritual, and belief. The more detailed answer addresses the many ways in which these behaviors offset aversive and undesirable brain and body states.

In previous chapters we noted two critical points. First, daily living has physiological and psychological costs to the brain and the body. It is *psycho-expensive*. Many of the inevitable features of daily living, beyond death and taxes, are stressful and aversive and are so in a continual stream of challenges. In response, people have devised numerous ways to offset these costs, such as holidays, retreats, physical exercise, sleep, entertainment, drugs, alcohol, and so on. Many of these are effective temporarily, but none offer lifelong assurance of stress reduction. Joggers get older, their knees weaken and they buy the largest containers of pain medication. Drugs and alcohol lose their luster and their kick. Spas become tiresome. And often too much entertainment may dull the brain.

OLD FRIENDS IN THE RIGHT PLACE
EQUAL BRAIN CHANGE

The *brainsoothing* that results from religious socialization works in a rather simple way. Believers gather at a location they consider sacred and where special rules of behavior apply. They see familiar and unfamiliar faces, but usually far more of the former, and even the new faces carry possible commitment to the common group and hence a better future. Familiar faces and their emotional expressions are reassuring and invite good thoughts toward those who are known. Other signals identify that they are members of our group. For some, acceptance of dress codes conveys the message. For others, it's words. And for yet others, it's behavior such as placing a scarf over one's head or drawing a cross in the air with one's hand. The congregants discuss their lives, their families, those who are well and those who are ill. The ambiguity and uncertainty that often accompany encounters with unfamiliar persons is largely gone. So is the uneasiness and hesitancy that can permeate moments in which the majority of faces are unfamiliar and their signals mixed and unclear. Perhaps it's these features of such gatherings that explain why most congregations are made up primarily of members familiar with each other.

The atmosphere in the sacred place is very positive. It is a haven and an oasis. Emotions are upbeat. Bodies are relaxed. There is a behavioral and emotional adhesiveness that is special and unique to such moments—a glue that facilitates basic human tendencies for generosity and tolerance. Those participating are in a place and involved in a moment they respect and trust, one from which they may leave wiser and better human beings.

And what a contrast to much of daily life . . . how simple but significant a change in state.

The brain is involved from the moment of arrival, if not before. Anticipating an event such as choosing an appropriate costume or seeing a familiar friend sets in motion those functional areas of the brain that are likely to be called into service once the event begins. For example, there is a cortical region consisting entirely of face-selecting cells, cells that inform its owner that "I know him or her" or "That's an unfamiliar face."[1] Much of this information is processed in the frontal cortex and stored both there and in the hippocampus for subsequent use. "Haven't I seen you somewhere before?" is not simply a pick-up line in sultry bars. It reflects a remarkable human skill at facial recognition, the brain mechanisms for which are rather suddenly becoming clear. We are neuroprogrammed for a cautious intimacy when it is possible and prudent.

But these are only selected examples of how the brain works and what it does at such moments. Much more is to come. For example, we've already noted that functional areas of observers' brains "mirror" much of the behavior and expressed emotions of those they observe.[2] That is, if a person is seen to be suffering, those parts of the observer's brain that would be involved in such suffering increase their activity too. Or if an observed person is happy, areas in the observer's brain active during moments of happiness mirror what is observed. This is largely a frontal lobe activity. The brain is encrypted to exchange such emotional information. Or, when an observer hears an action word, such as "kick," or "bow," or "sing," the same areas of the brain light up as when watching a person kick, bow, or sing.[3]

As we noted before, positive or negative words elicit activity in the amygdala, which is one of the brain's centers for processing emotion and influencing how information is interpreted.[4] All of this hints at how we understand and relate with others: our brains mind read by simulating those brain functions responsible for others' signals and behavior. It's simulation that explains why we cry when a friend tells us of his or her misfortune or feel joy when he or she talks of a promotion to chief of surgery. And, again, it's why people visit a movie house and watch images of the behavior of strangers on a cold screen and are moved as deeply and influentially as by signal events of their own lives.

Perhaps it's not surprising that we don't confine our mind-reading skills to humans. Most owners of dogs, cats, and horses would attest to this point as well. Their pets have brains too. And certainly people who spend much of their lives observing monkeys and apes come to feel this way. And not surprisingly, behaviors common among humans such as deception[5] are investigated and often discovered among nonhuman primates.[6] So also with those practicing animistic religions. Bears, wolves, and eagles are all objects of worship. Each is seen to have a brain, a personality, and an agenda. Animals behave and they emote, which is all that is required to put mind reading into action. More surprising perhaps is mind reading of inanimate objects. Farmers frequently talk to their tractors as if they were disgruntled employees—"She doesn't feel like plowing today"—and increasingly at the modern office one hears people mumbling to themselves and even others about the moods and stupidity of their computers and the Internet.

More surprising and often rather startling is the mind reading of statues and paintings. This is particularly true among Christians with depictions of Jesus on the cross and of expressive saints, however they are variously presented. The formula is clear: attribute a personality and emotions to the statue or painting and then mind read what one has attributed. Circular? Yes, but also often reassuring and evidently convincing to many. Even nonbelievers are not immune as they enter a church and experience the pained face of a statue of Christ. Much of the art of Christianity fits this explanation. One can easily imagine what was going through the brains of Michelangelo, Donatello, Cimabue, Leonardo, and the like.

Another point about religious socialization deserves mention. It's possible—perhaps more than possible—that humans are born with an egalitarian bias similar to nonhuman primates in which they refuse to participate in tasks if they become aware other animals are receiving greater rewards for completing the same task.[7] And it's far more than possible that the way in which social groups organize themselves hierarchically—featuring bosses, sergeants, teachers, and others—competes with and constrains this blissful bias toward equity, which is certainly not always achieved. It is a chronic source of tension. Call it "office politics" wherever it happens: the office, the backfield, the operating room, the zoning board, the chorus line.

Enter religion, where for believers as a group and for those moments in which they are with fellow believers, the hierarchies of the real world temporarily fade away. For a moment, the president of a company and a clerk from its mailroom are on equal footing. Both may serve meals to the poor.

Both may sing together in the choir. Both may bow their heads in unison at a moment of prayer. Both may wear headgear, such as a skullcap, recognizing their subordination to a higher power. And both may walk together to the parking lot. They recognize that tomorrow at work they will walk separately to very different places of power and privilege. But for an hour or two they are wholly equal, an equality certified by the greatest power in the universe.

Now back to signals and the brain. The positive signals given off by other people have immediate impact on the brain. *You are worthwhile. We share similar values. We acknowledge the same higher authority.* And, *we live by the same uniquely worthy rules.* These are just examples. Words are not essential to convey such messages. An extended look, an unusually friendly smile, an approach closer than is customary, or a willingness to listen attentively to a familiar story will do the trick just as well as words. "The medium is the message," as Marshall McCluhan, who knew his stuff because he was a lively Roman Catholic, famously noted.

Underpinning these signals are the chemical, neural, and muscular states of the body. They have characteristic patterns. They are readily readable. And they reveal a person's emotional and cognitive states.

SHOWTIME

The signals start arriving. The chemical and functional actions throughout the deepest recesses of the recipient's brain go to work. The play begins. The characters take their spots in the drama.

But before opening the curtain, a further word about science. Earlier we mentioned that science, and especially the kind of science being described here, is always changing, which is desirable. New findings and new interpretations of findings assure this. There is also another point directly applicable to the remainder of this and the next chapter. In complex physiological systems such as the brain—indeed, the brain is likely to be the body's most complex physiological system— the whole story of chemical and functional events can't be told even were it known. And it is not. What is known at present might fill up one hundred thousand single-spaced, small-type pages. Here we have selected only some of the high points for which there is compelling evidence and that are likely to remain significant ten and twenty years from now.

Among the main chemical players involved are serotonin, dopamine, and norepinephrine. These are all neurotransmitters or molecules involved in transmitting messages between neurons. And then there is oxytocin, which is both a hormone and a kind of neurotransmitter.

Serotonin is probably best known as the neurotransmitter that Prozac and other antidepressant drugs alter by elevating its functional levels in the brain. In turn, this often ameliorates the signs and the symptoms of depression. But long before these drugs arrived in the medical market, it was known that among certain Old World primates there were unusually high levels of serotonin in the brains of animals of high social status.[8] Conversely, low levels are found among animals of low social status. And when animals change status, their serotonin levels change. Of course, it is hardly surprising that when things change, things change. No kidding. Yet the specific patterns of secretion of serotonin are fascinating and definable. They tell us vital things about equality and dominance and promise to reveal still more as scientific work goes on.

How these serotonin findings came about not only is interesting, but it is also illustrative of how sometimes science works unsystematically and by chance.

Wander back to the early 1980s, a period in which studies

of relationships between behavior and brain physiology were in their infancy. At that time, one of us, McGuire, was director of a nonhuman primate laboratory at the UCLA Medical School that consisted of two dozen stable groups of Old World vervet (*Cercopithicus aethiops*) monkeys, which included dominant and subordinate males and females and their offspring. Studies were focused primarily on the behavioral features of dominant and subordinate males, the moment-to-moment operations of their hierarchies, and the conditions under which dominant animals would lose their status and subordinate animals would become dominant—actually a rare event for this species, as it happens approximately every two years.

At the time, Old World, nonhuman primates, such as rhesus and vervet monkeys, were preferred to New World monkeys as study animals. This was because it was assumed that findings would more accurately extrapolate to humans and because much of the research was funded from medical sources. There was clear reason for concern about what this meant for people. Little did we know.

During this period, a postdoctoral fellow came to work in the laboratory. He was well versed in the mechanics of observing nonhuman primates but had no experience in chemistry. To further his training, he was assigned to a nearby chemistry laboratory to learn some of the techniques of assaying biological tissues.

One of the most efficient ways to understand chemical events in the brain while not sacrificing animals is to examine their cerebrospinal fluid. This is a fluid that without serious danger can be extracted at the top of an animal's spine. Cerebrospinal fluid is essentially the brain's garbage system in that it is made up in part by the by-products from the breakup of the brain's neurotransmitters. By chance it was decided to obtain and assay 5-hydroxyindolacetic acid—5-HIAA for short—which is the primary breakup product of serotonin.

Training began. Samples of cerebrospinal fluid were extracted from a number of dominant and subordinate vervet males. The chemical analysis for 5-HIAA commenced.

Much to the surprise of everyone and the chagrin of the trainee, the findings showed a near two-to-one ratio difference in the amount of 5-HIAA in the cerebrospinal fluid of dominant males compared to subordinate males. Big shots produced big numbers. What that two-to-one ratio suggested, and what subsequent studies came to show, was that the serotonin system in the brains of dominant males was twice as active as it was in the brains of subordinate males. (Subsequent studies did not reveal similar differences among dominant and subordinate female vervets.)

This was extraordinary. Most findings in natural science provide results along a normal curve of response. Relatively small differences become critically important. But these differences were massive and seemed to be like none other. "The findings must be in error" and "They're an artifact" were the first two sentences anyone heard. There was nothing in the scientific literature of the time and nothing among the hypotheses that were responsible for the research at the laboratory that hinted even slightly that such differences might be found. It was, after all, well known at the time that many of the familiar body chemicals of nonhuman primates and humans, such as glucose, sodium, potassium, and so on, were essentially the same among high- and low-status animals and individuals. So, "It's an artifact of something unrelated to dominance" was the consensus conclusion for the day.

Things then took a turn in another direction. We had to assume that such dramatic findings were in error. Now it became essential to find the source of that error. Otherwise, all future assays would be suspect.

Our stubborn decision was to repeat the study until the source or sources of the error were detected. There were many possibilities. It could be the time when the cerebrospinal fluid was removed from the animals. For example, cortisol was known to be at higher levels in the morning than later in the day. Or it could be how the cerebrospinal fluid was stored on its way to the laboratory for chemical analysis. Or it could be in the laboratory chemicals used in the analysis itself. Or it could be other unsuspected reasons. It was essential to find out.

The original study was repeated six times. Each time, we carefully monitored each step. We systematically altered the times the cerebrospinal fluid was drawn. We paid precise attention to the transportation of samples from the monkeys to the chemical laboratory where they would be analyzed. In the chemical laboratory, we employed different people to analyze the fluid for 5-HIAA.

But the two-to-one ratio held up! This was in spite of good reason to suspect that it wouldn't, in spite of multiple attempts to disprove it. It was intensely thrilling and rewarding—hair-standing-up-on-the-neck time. The door had opened not only to a truly novel and totally unexpected finding, but also to a novel research strategy, which no one had thought of before. For example, 5-HIAA levels could be followed when animals changed status. Or, animals of low and high status could be removed from their groups, temporarily socially isolated, and the effects of isolation on 5-HIAA studied. We had a new marker at the core of social and organic life.

People who are not scientists or involved in research may find it difficult to grasp how findings like the two-to-one ratio can have an impact on the lives of those participating in the research. No one had expected, perhaps even thought of, the two-to-one possibility. The core of science is discovery. We had discovered something. And suddenly everyone knew.

It's not quite like finding an answer to a question on which one has worked for months, if not years. This was the case, for example, with Crick and Watson and others who decoded the structure of DNA. Nor is it like winning a substantial sum from a state lottery because, after all, one bought the ticket and invested in the tiny chance there would be a desired outcome.

Rather, the two-to-one ratio was more like a revelation, a sudden jolting realization in the brain and the heart that we had overlooked a critical and defining feature of precisely what we were studying.

What we had been thinking for years past required instant and significant revision. What we had to and wanted to do tomorrow in the laboratory demanded a new format. The next week in the laboratory we developed not only a new format, but in addition the lives of the researchers and what they did swiftly changed.

Within moments it was clear that now there was a new paradigm. Social behavior could affect the biochemistry of the brain. The skin was a transmitter—both ways. Behavior could affect the brain significantly, and knowably.

There were huge, unforeseen, and unimagined implications, which needed to be sorted out and pursued. Within moments, it was clear that the academic careers of literally all those involved in the research were ripped away from their expected trajectories. They transformed in a few days into entirely new and still uncharted directions.

This was just the part of the story that took place in the laboratory itself. But change excites the world of science just as it does the public. This excitement was detonated by a major story of the findings that appeared in the *New York Times*. The telephone began ringing ceaselessly from all time zones. Invitations to speak about the findings came from academic centers around the world. Requests to consult in similar laboratories quickly appeared. There were resplendent invitations to the senior investigator for well-paid and well-funded professorships at Ivy League universities. Reporters from myriad newspapers demanded interviews. Scientists from distant continents came to work in the laboratory. Students from various universities offered their services if they could be part of the colorful new agenda.

Throughout all this, we continued to wonder if the findings were a matter of chance. Or perhaps the research group was headed for the findings all along and didn't realize it. Or was there another reason?

But without question, we had discovered and isolated an unexpected route of influence between what complex animals did in their behavior and what their bodies and brains did as organic systems. We had reversed the normal path we thought we understood, which was that experience began in the body and affected the community.

No. It was the other way around.

And not only that: this was a feel-good process. Serotonin made animals, and presumably people, feel well, better, good, okay, but not depressed and disconsolate. We had found something genuinely new about God's brain. Yes, God has a brain, as you will soon see. We had discovered that the brain could self-soothe with its own juices if it were owned by a socially convivial person able to engage warmly but assertively with other people.

It's no surprise that it then became intertwined with Prozac and other drugs, which have often decisive impacts on social behavior.

THE BOSS'S CHEMICAL SET

What did we go on to learn about the brand-new link between serotonin and status? And what could this mean for the practice of medicine? And what could this mean for people?

Serotonin levels positively correlate with the number of submissive displays an animal receives. This is a vital behavior by which subordinate animals signal to dominant animals that they recognize the dominant animal as the group's leader. It certifies that the signaler is accepting the group's hierarchy and rules about behavior. High-status animals receive far more submissive displays than low-status animals. Remove a dominant animal from its group, which of course means that it ceases receiving submissive displays from subordinate animals, and its serotonin levels decline. No status juice. Put him back in his group and, should he regain dominant status, his serotonin levels increase.

The human equivalents of submissive displays from others—an extra-friendly smile, an extended handshake, many phone messages from celebrated people, special delivery letters, especially rapt listeners, and so on—signal that the recipient is an important person, necessary to the group, and respected. Said another way, the signaler is declaring, "You are important enough to merit my attention and good will." Such signals elevate a recipient's sense of status. And that sense is associated with feeling good, physically and mentally comfortable, responsible, socially important, and in charge. And this may help explain why men who retire die twice as often in the first year of inactivity as in the next—sudden serotonin deficit disorder?

This is just the opposite of how people feel who are low on the social-status ladder, depressed, or at sundown are rundown after a stressful day. The more positive social signals one receives from others, the better and the more necessary and vital one feels. This is due largely to the chemical changes in the brain. Most people just can't get enough of a good thing.

The likelihood of receiving and sending such signals takes a quantum jump at religious gatherings, compared, say, to work or many family interactions. We repeat that one of the two major reasons for this jump is the temporary disappearance of hierarchy as an obligatory basis of relating among members of religious groups. The other reason is positive signals. Positive signals exert their effect on serotonin levels even in the presence of hierarchy. We are in *Shangri-la*, if only temporarily. We share a common resistance to threats from nonmembers. As analysts, we can use this perspective to see churches, among other things, as serotonin foundries. Even outdoors, as in Vatican Square at Easter and Christmas, the collective ritual becomes a private reality made concrete by the potent, silent chemistry of the brain.

Let's digress a moment. There might be those who doubt that communications from others alter the brain, its chemical profile, and its functional activity and efficiency. In that case, recall times in which you have been the target of negative signals, such as serious criticisms of your work or partner or appearance, and how you felt. Or, better, recall when you have received a traffic ticket with a big fine that you didn't deserve or you were accused of a serious breach of trust when you were innocent. Or recall a moment of unexpected good fortune, such as a letter advising you of fellowship in the National Science Foundation or a certificate awarding you the national gimmickry medal for 2009. And then recall that such happenings are only communications, words and facial expressions from others. Apart from the cash extracted by the fine, the other negatives were largely social and reputational. But the accountant-brain calculated their depressing value.

Neurophysiology, not economics, may be the real dismal science.

For technical and ethical reasons it remains difficult to precisely measure serotonin levels in the brains of humans. But literally all available evidence is consistent with the idea that high status and elevated serotonin levels go hand in hand. For example, tryptophan, an amino acid, is a nutrient and a molecule essential for the synthesis of serotonin. Administering it increases its levels in the brain, leading to a decrease in quarrelsome behavior and an increase in dominance behavior among normal humans.[9] The administration of Prozac to psychologically and physically normal individuals leads to a decline in their hostility and negative affect and it increases comfort with social affiliation.[10] In a way, it may almost allow them to feel they have had the promotion in rank in real reality associated with serotonin. Serotonin is known to modulate behavioral reactions to unfairness[11] and influence self-regulation.[12] The fact that more women than men use Prozac and similar drugs suggests they need the promotion more than men in an occupational world more hostile to women and more difficult for them to navigate.

These and related findings offer insights into the aversive status-reducing events and communications that often compose the menu for the day. They are the source of stress. They most likely cause reduction in serotonin. Other studies show that a depletion of serotonin in the prefrontal cortex—the thinking part of the brain—leads to cognitive inflexibility involving stubbornness, defensiveness, negativity, and resistance to changing one's beliefs.[13]

Low-serotonin suave management consultants and bosses beware. This is very likely to be far more important for the conduct of human affairs than authoritarian managers can ever realize. But their subordinates know it all too well. And it is very provocative that the synthesis of serotonin in the brains of males far exceeds that of females[14]—recall that 5-HIAA differences were not found among female vervets. Perhaps males need to secrete more serotonin than females to control their competitive urges and play well with others. In fact, this is a huge subject about which we know far too little. More insight into the process could have immense practical consequences for the conduct of organizations.

Interestingly, even though the data about the links between serotonin and social status have been known for over twenty years, the precise mechanisms by which signals from other people either raise or lower serotonin remain a mystery. Obviously, others' positive signals need to be visualized, heard, and/or felt. So the visual, auditory, and touch systems of the brain will be involved. But so too will be other functional systems and centers. Key among these is likely to be the amygdala because of its known sensitivity and critical function in processing positive and negative social information and its involvement regarding optimism about future events.[15] As well, the prefrontal lobes are involved because of their pivotal part in perceiving and evaluating the minds of others.

Still another point about this fabled juice, serotonin, deserves discussion. Unlike the rapid increase in adrenalin that takes place when one is frightened or in dopamine when one anticipates that pleasure is just around the corner, serotonin levels don't change rapidly in response to positive social signals. It appears to be a careful and conservative neurotransmitter. There are temporary increases that, for example, occur during certain types of meditation, but they fade quickly. Generally, at least a week of repeated positive social signals is required to elevate basic levels. One doesn't recover from low social status in fifteen minutes and, as with Citizen Kane, it may take a lifetime. The requirement of repeated positive social signals predictably draws people to certain environments and company.

And then there is the often-complicating factor that during the same day one's social status in the real world may range from high to low depending on circumstances and relationships—for example, high in one's family, low at the office, and medium on the bowling team.

One absolutely certain measure that change has occurred is that people feel good when they are alone. They feel good even away from the company of the others who have been sending positive signals. It is also desirable that the over-used and aerated term *role model*, with its implication that others matter more than the self, should be abandoned in favor of a perception of personal well-being founded in a benign personal neurophysiological cocktail. The happy hour drink should be serotonin and soda. Perhaps this is an explanation for why many believers participate in religious services and related activities such as church dinners, prayer meetings, and social support projects four or five times per week during down periods. It is also worth recalling the years of contrite self-remedy of disgraced persons such as John Profumo, who as English defense minister engaged in trysts with a woman also beguiling a Russian diplomat, and Charles Colson, who participated in the Watergate shenanigans, to gain a sense of the depth of neurophysiological loss and what is necessary to reverse it.

What happens with serotonin also appears to happen with norepinephrine and dopamine. Norepinephrine is known to increase in the brain during positive social encounters. Dopamine activity is known to increase in anticipation of pleasure and its desired effects.[16] Part of religious socialization is anticipation of the pleasure of social interaction. No doubt such vivid reminders as Christmas and Easter decorations prepare the ground for what will happen in the skull during the main events.

OXYTOCIN INTERMISSION

Let's return to our theatrical metaphor. If this book were a play, it would be time for an intermission. At this point we are surely fatigued from confronting and trying to understand some of the most complex and dynamic processes within our body. The challenge is rather like visualizing nothing less than the electrical wiring of a Boeing 747. Not only are newly identified neurotransmitters revealed on an almost weekly basis, but there remain some centrally important ones we have yet to face.

One of the most interesting and supple newcomers to the cast of characters is oxytocin. This is best thought of as a multitasking chemical, which busies itself with, among other things, stimulating reproduction and enhancing sociable moods. It acts as a hormone when it induces labor and lactation in women. Its neighborhood is heavy-duty core behavior.

It serves as a neurotransmitter in particular brain areas associated with emotion and social behavior, such as the amygdala and the nucleus accumbens. It contributes decisively to social attachment.[17] Among lower animals it appears to reduce resistance to having other animals nearby. Otherwise, they predictably space themselves away from each other in ways that reflect their status and dominance relationships. And its levels increase in social interactions where people trust one another. Elevated levels of oxytocin improve one's capacity to read others' minds and also lead to a warm feeling in one's body.

Where there is chemical change, there is also change at selected functional centers. It is no surprise that because of individual differences, such changes often have interesting features. For example, as we have noted, a person's amygdala responds differently to the happy faces of other people as a function of the personality type of its owner.[18] Extroverted people show much greater amygdala activation to positive social signals than people do who can be identified as introverts.

As a result, introverted people appear to require a markedly greater amount of positive social input to achieve the same degree of change in their brains as extroverts do. They need more Worcestershire sauce in their cocktail of interaction. Each person not only looks different from all other people, but also the chemical look of what goes on within the skull differs too.

A similar point applies to experience. A person who has just been convincingly rejected by someone very important to him is unlikely to respond eagerly to positive signals from other people. At other times, when things are going well, a positive response occurs easily.

So where are we in this play?

We have been trying to define what works for the brain the way vigorous exercise grooms the body. The time will surely come when restoring brain health will be as obviously a feature of will and arrangement as exercise is of physical robustness. This will depend as much on skills at looking inward to the brain as identifying outward causes, which is now done routinely in the social and physical environment.

The picture—actually a rather encouraging one—begins to emerge. Optimizing the brain's chemical profile and functional operation is the principal if not the only basis for *brainsoothing* a stressed brain. The process is facilitated by positive social input that in part offsets the brain chemistry, which results from the aversive and self-deflating effects of daily life. And these produce desired and desirable consequences, such as reduced anxiety and fatigue and noticeably improved body comfort. Said differently: on average, religious socialization leads to predictable, desirable outcomes among believers. Were this not the case, participation would decline. And where participation has declined, such as in large swathes of Western Europe, we may surmise that the contributions to *brainsoothing* of the available religions are less rewarding than those available in secular realms. Or, even if they're not, the social barriers to religious observance are not seen as worth surmounting.

There has to be a strategy here, to seek out and engage in social encounters with a high probability of a positive result. One's brain has to recognize the connection between positive social input and decreased symptoms of irritation and stress.

The body has to be commandeered to transport its brain to just the right social environment at the right time. Get me to the church on time. Such encounters are not necessarily available at home or at work. They may be absent, for example, at town meetings about zoning or business conferences where people may be involved in contentious debate—in fact, the main function of such meetings may be to house contention.

But these are rather more predictable in religious settings. With rare exceptions religions treat social interactions with great dignity and equality. While a serotonin fix is not guaranteed, it is a good bet there'll be enough of one, and at low cost too, to justify the drive or walk to the mosque or church or temple. The experience is rather like putting on a fine new autumn coat or a making a springtime visit to a buzzy botanical garden.

But we've spent long enough on a complex subject. Let's bring this chapter to a close. Let's turn to the other two signature features of religions: rituals and beliefs.

Chapter 9
PUZZLES, ANSWERS, AND MORE PUZZLES

If religious socialization is one part of a religion's antistress holy trinity, rituals and beliefs compose the other two parts. They are as central and as critical to bringing about chemical changes in the brain as positive social input. Serotonin is again part of the story.[1] But now it is only part. Other brain chemicals and functional centers as well as a host of other changes in the body take their important moments on center stage.

We have, please, to bear in mind that what is under review here is nothing less than an explanation of the mysteries and realities of religion. And this must be accomplished with neither partisan piety nor confident hostility. Our challenge is to explore how they are integrated into the operations of the human brain.

We have no better explanation for the origination of religion than as a product of the ever-active human brain. We may appear to discuss rather casually particular brain juices and squishy bits. But we are sharply aware that these discussions are neither trivial nor arbitrary.

Instead they direct us to the heart of the problem of understanding religion. They offer an account of what actually happens in human heads to make religion possible and to form it and to sustain it. This is clearly important for scientists and people trying to live and function in the world.

RITUALS

There is nothing unfamiliar about rituals, of course, by definition. They are everywhere, every day, every moment: what time school starts, when and how to salute the flag, who sits down first at the dinner table, how one manages a four-way stop-sign intersection, where one stands in line at the grocery store or a movie queue, and how one is seated at weddings.

These are all behaviors. But they are rituals too. Rituals are rules—usually not laws—people live by, ways we behave in both public and private. They are integral parts of everyday behavior. They are like the joints of the skeleton. Groups and societies rely on them to function efficiently. They signal a member's savvy about what to do when and even why. They reveal mental values and states to others. They are a secret code everyone knows, much like informal laws.

The strong conclusion presents itself that carrying them out usually makes people feel comfortable and safe, at least for the moment. And they deliver the message that the individual knows the rules and therefore probably *belongs*. Rituals are like linguistic accents—they mark membership as clearly as they permit communication.

If rituals are to be efficient and likely to endure, they have to accommodate human predispositions that have a genetic basis. For example, courtship rituals among teenagers, reproduction rituals among the families of new offspring, and sickness and death rituals among loved ones are all knots of behavior driven by the inexorable features of the life cycle.

Same life cycle anywhere. But different rituals everywhere. It's likely that most rituals are learned in a specific group. Live in country X and you learn and abide by one set of rituals. Live in countries Y or Z and there will be other sets. You are likely to be uncomfortable with someone else's rituals. You may have to develop a wary stranger's pose. But you will likely adapt eventually, unlike the legendary British imperial colonial officer who dressed in an ivory dinner jacket when he dined alone in his remote station in inland India or steamy Nigeria.

From afar, sacred rituals appear much the same in their timing and solemnity of enactment. They often have the same function but different forms across groups. Every major religion has its list of rituals. For example, in Judaism, Islamism, and Hinduism literally every important milestone from birth to death—maturation points, marriage, bearing offspring, attaining official membership—is ritualized. So also with the details of religious conduct. Praying, singing, chanting, moments of silence, who leads in processions, how sacred texts are placed and held, and the order of prayers and blessings are all examples. Christianity and Confucianism follow closely with baptism, liturgical observances, holy water, the celebration of sacraments. (Even witches have rituals. Successful sorceresses must go through a variety of weird speaking and behavioral exercises to certify they possess the fright profile to be able to efficiently alarm others with the sentiments of the devil.)

Of all the world's major religions, only Buddhism offers a seeming exception, yet not completely: after all, there is formal and predictable behavior, which links teacher and student, and many engage in ritualistic behavior when they pray to statues of the Buddha for enlightenment. Their private occasions are the essential and core rituals of the faith. Other groups generate more public sacred rituals that produce the same *brainsoothing* effects as religious socialization. People socialize at baptisms, marriages, funerals, and before and after prayers at meals.

It is perhaps redundant but nonetheless necessary to point out that believers are acutely aware of even minor violations of appropriate practice. Imagine the fuss if a veil is removed in a mosque, a woman attends High Mass in a bikini, or a geek tip-taps on a chirping laptop computer during the recitation of the Lord's Prayer during a cathedral service.

Rituals are often unexpectedly colorful, complex, and varied. But there is an element common to them all: the participants' subordination to a higher authority, however imprecisely it may be defined. For each individual and for the group as a whole there is clear subordination ideologically during prayer, physically and mentally during meditation, mentally and spiritually during confession, and operationally in what one does and does not do. Rituals help define what one is to eat, how one behaves in sacred settings and with fellow believers, and how one devotes one's time and resources to religious activities. Objects and aesthetic initiatives created for rituals—for example, crucifixes, painting styles, or specialized head coverings—may carry over to domestic décor and even clothing styles.[2]

At its essence, this is subordination, the straightforward stuff. It is predicated on the presence of a deity or higher power.[3] No matter what form it takes, it is a very different type of subordination than, for example, remaining politely quiet while the minister preaches. The higher authority or power can't be visualized, just imagined, just inferred. As noted, producing images of the God of the Jews is strictly forbidden—surely this protects the deity.

However, ministers can be seen, touched, and heard. Their handshake at the doorway after the service is customary and valued. The authority of cardinals or imams or rabbis is in principle no different than the authority of a boss, a sergeant, or a president, although some cardinals, charismatic preachers, and so on might like members of their flock to believe otherwise. (The limited notion of papal infallibility on doctrinal matters and the option of senior imams to issue fatwas are breaches—and controversial ones—of this principle.) Submission to a higher authority differs in part because real live ministers change, die, and have personalities. Different popes conduct their work differently, often to the consternation of many believers. But the supreme authorities live on, minimally changed and permanently untouchable.

There are rituals connected with the rituals described above that require a very special behavior. These can be usefully highlighted. They usually require specific bodily or mental outputs. They often have to be performed in a designated location such as a mosque, temple, or church. They are repeated literally daily, if not more often, five times in the case of orthodox Muslims. They are patterned and explicit. Each behavioral element can claim theological or at least customary justification. Praying and meditation are examples. They range from the very simple, such as a short prayer, to the highly complex. For example, there is a lengthy Taoist meditation ritual that purports to increase internal sources of energy.[4] Here is the recipe:

- Begin the meditation by taking up the *wu chi* (emptiness) stance.
- Keep the body erect but not stiffly upright.
- Stand with your legs shoulder-width apart and slightly bent at the knee, with your feet parallel.
- Your breathing should be steady and even. Breathe out for a little longer than you breathe in.
- When you are comfortable in the above stance, become aware as you breathe in through your nose of your *tan tien* (point of power in the abdomen) and the expansion of your abdomen.

- As you breathe out, either through your nose or mouth, be aware of the contraction of the abdomen.
- Now focus your mind on the area of the lower *tan tien*, where the *chibai* (reservoir of power) or "sea of chi" is situated.
- As you breath in, imagine fresh *chi* (energy) from the universe entering your body and flowing into the *chibai*.
- As you breath out, imagine the now-depleted energy leaving your body and returning to the Source (the universe), where it will be renewed.
- If you can, continue to do this for about ten minutes and feel the power of *chi* as you store it in your body.

These rituals are of special interest. They yield hard evidence that they change the brain's chemical profile and function and reduce the aversive effects of stress—they seem to work! For example, meditation or any type of intense spiritual contemplation is usually associated with decreases in oxygen consumption and heart rate.[5] They are followed by decreases in important body chemicals such as lactate, cortisol, and ACTH.[6] There are decreases in blood pressure and alpha wave activity in the brain.[7] There are increases in galvanic skin response[8] and urinary 5-HIAA, which, remember, is the metabolic product of serotonin.[9] And given our interest in the brain, it is intriguing that there is enhanced cognitive focus.[10] The prayer maker feels less anxious, more relaxed, and (for better or worse) a greater sense of personal efficacy. It's almost as if the ritualizer can control his or her brain in the way a trained gymnast can stand on one hand for a surprising period of time. And why not?

TIME FOR A COFFEE BREAK

For the brain to try to understand itself is difficult. As we've said before, it's rather like asking a centipede, "Hey, just how do you walk?" So also with pondering one's own brain and what it's up to and why. The challenge becomes greater but also more significant when what happens inside the head is linked to what happens outside the body. We have sought here to suggest that very important and widely practiced human rituals are directly and knowably linked to the operations of the brain.

Many different rituals are performed in religious worlds. Different findings about internal brain function are to be expected when different rituals are under study.

That's just what we find. For example, there has been fMRI analysis of traditional meditation. During such meditation, there is increased activation observed in a number of functional areas of the brain, such as the dorsolateral, prefrontal, and parietal cortices; hippocampus; temporal lobe; and so on. This is a critically important insight. It means that there is increased activity in structures involved in focusing attention and control of the autonomic nervous system. And this is a brain system that affects the body and is closely associated with stress and anxiety.

Other studies show that Zen-type meditation increases the activity of the frontal lobe and basal ganglia. It decreases activity in the gyrus occipitalis.[11] Yet other studies suggest that spiritual meditation is more effective in raising one's threshold to pain compared to secular meditation.[12] It's better medicine. People have been performing these rituals for centuries. At last we know what are the actual and factual mechanisms of the brain that support them. So when the fruit and vegetable seller at a street corner kiosk bends down deeply in prayer to touch his forehead to the cold pavement, something is presumably happening inside his body just as a remarkable behavior is outside, in full view.

Positive environmental cues appear also to activate brain regions associated with reward.[13] These cues may find their origins internally or externally. This suggests that people can temporarily change their brain chemistry with certain thoughts and prayers.[14] Compare how you feel after a period of mean envious thoughts about your elegantly and unfairly fortunate enemies with pleasant thoughts about lazy vacations in tasty, inexpensive Italian hill towns.

As mentioned, we also know that the amygdala is directly involved in positive and negative responses to visual images and optimism.[15] Perhaps this begins to explain why there is such affluence of visual displays of higher religious powers in many churches and similar buildings of other faiths. It is also likely to begin to reveal why people are willing and eager to wear amulets and other symbols of belief and attachment, such as crucifixes and Stars of David, on their bodies. Believers focus on objects because they are sufficient symbolic ways of affecting the brain in a reassuring and energizing manner. They are crisp and to the point, like words.[16]

But like many of the positive effects of religious socialization, those of rituals are short lived.

Sustaining their desired effects requires repetition, just as one must go to the spa or exercise vigorously more than once every six months to achieve maximum effects or even any at all. Much of the reason the effect is short lived follows from the fact that after a ritual, it is usually necessary to return to the glitch-filled real world. There our ever-active social brains leave us highly vulnerable once again to signals from folks unlikely to be as committed to our contentment as are the managers of the rituals.

Nonetheless, there are puzzling differences regarding rituals. For example, why is it or how did it happen that Muslims have come to pray five times a day, that some travel ten thousand miles to perform the sacred hajj, that some Christian fundamentalists pray for two hours a day on their hands and knees in their bedrooms, that some Catholics attend Mass twice a day, and some Buddhists meditate for two or more hours per day, while others spend but a few minutes per day or week or month in these activities? Is this the specific amount of effort required to successfully *brainsoothe*? It does not appear to be the case. It's enough to try the patience of theologians.

Something else must be at work. One possibility is an attempt to extend the desirable *brainsoothing* effects of ritual, much as one's brain replays music following an emotionally moving concert or relives moments of pleasure following an intimate interlude. Serious worshipers offer another possibility: at moments during rituals, "I feel as one with my God . . . nothing in daily life is quite so satisfying." Such experiences are powerful and maintained in memory. Surely memories of such concrete experience support people when they consider the plausibility and good sense of adhering to religions whose logic and empirical validation are often frail and scarce.

One additional tantalizing finding deserves our attention. Numerous studies show how changes in an adult animal's physical environment transforms sensory maps in the cortex of animals.[17] So too with humans.[18] There is clear evidence of brain plasticity. Within limits, the circuitry and functions of the brain may be modified. But it is by no means clear exactly how this happens. And this is surely one reason Buddhists and other serious mediators are so attracted to meditation. They appear to do nothing. Yet they are changed.

The brain is part of the body and of course affects behavior. Rituals are behaviors with special, often specific, impacts on the brain. Though they are often cloaked in historical haze and imprecise theology, rituals are nonetheless suddenly knowable as brain processes, which in final analysis they are. There is a road map of *brainsoothing*—an inner global positioning system. Someday we may be able to project it on a small screen.

RELIGIOUS BELIEF

Religious belief is the third signature feature of religion. It is what is written in sacred texts, what is said by members of the cloth, what is passed on among believers, and what a person believes, whatever its source.

Please recall that the first two signatures are ritual and socialization. Like religious socialization and rituals, beliefs *brainsoothe*. They chart the unknown and the future. They provide a proposed menu for the great permanent feast of the afterlife. They answer questions. They serve as a formula. They offer a strategy for how to live on any continent until life somewhere begins after life on earth.

The prospect of complete nothingness after death appears to be bewildering and unendurable to many people. So an antidote arises. This permits everyone to abide by a personal contract one writes for himself to deal with what he knows he should do in the context of what he fears he will do. Religious beliefs boost self-esteem and the sense that one has a legitimate place in this and even other worlds.[19]

The belief antidote demands and elicits remarkable, even astonishing boldness and invention. Ambiguity and uncertainty are reduced about matters of life, death, the soul, eternity, and the like. It is hectic, like doing calculus during a rock concert. Yet it seems people somehow manage. Furthermore, it appears that everyone contemplating heaven or hell is an intrepid explorer and has been issued a guidebook, however imprecise it may be.

BELIEFS AND AMBIGUITY

Beliefs reduce ambiguity and uncertainty uniquely well. There are two basic types of ambiguity and uncertainty.[20]

There are those associated with real life. Then there are those that result from human imagination about events and facts that can't be proven.

Normally the brain avoids ambiguity and uncertainty. It likes answers, concreteness, and predictability. It prizes the material with which to make the right decisions. To reduce uncertainty, people read books, peruse newspapers, ask questions, watch the news, draw statistical graphs of the investment histories of companies, and concoct authoritative historical stories about why recessions begin and end. As we've noted somewhat haplessly, an astonishing number of people read horoscopes and even seek personal consultations with astrologers of good reputation if not accuracy.

But there remain real-life situations in which outcomes are unavoidably ambiguous and uncertain, such as the outcome of a serious medical operation or of going to battle on the front line in a war. Sometimes the odds are dramatically uncomfortable and hover between ambiguity and risk.

But for the brain there appears to be a significant difference between ambiguity and risk. It responds differently to each. We can observe it—ambiguous information lights up significantly more functional areas of the brain than information about risk. This seems very strange that ambiguity is harder to cope with than risk. Risk is potentially very painful but easier to evaluate. A clearly dangerous risk is more easily absorbed than fuzzy ambiguity, even if it's menacing. Remembering that the brain evolved to act, not to think—to act, *not* to think—suggests why this is so.

The brain is also very active chemically. Busy busy. For example, the activity and amount of actylcholine in the brain correlates with the degree of expected uncertainty.[21] Expected uncertainty is when you don't know the outcome, but the event is constrained by time. For example, at the end of three hours someone will win a football game. The results of the ten-million-Euro lottery will be announced on Tuesday or the Illinois Powerball number on Friday. By a specific date an election will be held. The activity of the brain juice norepinephrine relates sharply to the degree of unexpected uncertainty. No one confidently knows just why, but that's the fact.

Unexpected uncertainty is about events when the outcome is completely known and the date of the event is unknown—for example, the date and location of a person's death.

In contrast, total uncertainty is when it is not known if an event will occur at all and, should it occur—for example, should there be a hell—when it will occur, and what, please, might be one's date of arrival there? Does it matter? No. Does one want to know? Yes. Total uncertainty is far more disturbing and painful than the less definite forms. Atonal music is less compelling than music played in the familiar key of C.

But how do religious beliefs decrease uncertainty and ambiguity? Primarily they do so by providing explanations and stories that must be believed (held to be true) to have their effect. No ambiguity. No backsliding. No fuzziness.

The exact details of how and by what processes people come to believe some things and not others remain a mystery. But something is known. For example, the brain filters information when it is interpreting and reconstructing the world. Remember what you missed in the first showing when you view a movie for the second time? It has to use prior knowledge for such reconstructions because information arrives at the brain in bits and pieces, never as a complete story with all the details in place. Such knowledge may be flawed. But even so, there may be few negative consequences of believing even strange imaginings.

At times flawed knowledge has its interesting, if not amusing, features. We have been discussing our nonhuman primate cousins, and a few further examples involving them may teach us something.

One of us, McGuire, worked for years studying primates on the islands of the Eastern Caribbean, in South America, and in Africa. We can draw from there three helpful examples of flawed but largely harmless beliefs.

One of the interesting, even bizarre, but unsatisfactorily explained facts of nonhuman primate research is that the skeletons of recently passed nonhuman primates are rarely—extremely rarely!—found in the wild. This seems surprising, if not incredible, given all the specimens of human ancestors of thousands of years ago that have been found on every continent. Still, it is a mystery and a fact. Local people and researchers literally never find bones despite years of familiarity in areas rich in animals.

What can possibly be credible explanations for this forensic puzzle? In both the Eastern Caribbean and West Africa many locals explain this situation with the belief that animals bury their dead, despite the fact that no one has ever—ever!—observed them doing so. The explanation is consistent with the evidence, of which there is none.

In many areas nonhuman primates are considered pests because they savage crops such as corn, fruit, and nuts. Often they are actively trapped and hunted and killed. The view of local people is similar to what citizens of large cities think of Norway Rats. They are useless pests, perhaps the source of diseases and filth. When in the 1970s researchers from American universities arrived in the Eastern Caribbean to study the behavior of the resident monkeys, many local residents rejected the idea that the researchers' motives were scientific. What were these strange visitors doing? Eventually, they inventively solved their explanatory problem very brightly.

They came to believe in those palmy, if strenuous, pre-Viagra years that American males were suffering from impotence. Therefore, the researchers must surely be on the islands to capture monkeys and remove their testicles. Then they would send them to laboratories in the United States, where they would be ground up, mixed with an elixir, and peddled to impotent American males. They did not presumably know of a distinguished user of some neomedication, namely, William Butler Yeats, who apparently was a devotee of a similar and earlier trendy rebuke to age. Of course, were this true, it would be a compelling example of the kind of primate scaffolding we discussed earlier.[22]

There was even an interesting subplot, which was that the most needy males were professional football players who were thought to endure severe impotence. But as Dr. Freud would have surmised, our West Indian analysts concluded that players sought to convince their audience through their rough play that they were real males. And for this they surely needed their monkey balls cocktails.

For unclear reasons in areas of West Africa, certain villages would occasionally experience an increase in the number of nonhuman primates in or near their villages. In a remote area where one of us spent considerable time, locals offered an elegant explanation: "The monkeys are multiplying and planning to take over the world." They are involved in "domesticating humans" so that they can live in and exploit that human world. And humans who believe in the "Monkey God"—a King Kong–like figure—would be allowed to live. Religion, yes or no?

This all happens in the jellyfishy tissue of the brain. The process of believing takes place largely in the frontal lobes and the posterior medial frontal cortex, squishy brainbits more critical in decision making and dealing with uncertainty.[23] Yet this may not always be the case at certain religious moments. For example, there are reports of a decrease in frontal lobe activity when highly religious subjects believe that "the spirit of God" is moving through them and "controlling them to speak." Still other studies show that believers have greater electrical activity in the right hemisphere of their brains compared to nonbelievers.

The neurophysiologist Michael Gazzaniga has postulated the presence of a neocortical "storyteller." While this storyteller is often wrong, its story may have survival value. Generally the brain prefers happy stories that provide details of desirable outcomes for certain behaviors. Many of the stories religions provide fit this category and therefore may be useful. What's wrong with believing in an afterlife?

Factually, the illusions of religion result from the brain's actions to fill in the incompleteness of experience. They provide some substance to the unknown. And, of course, there is a large and heaving raft of terrifying stories that depict events if rules are not followed and if subordination to the gods and their representatives is insincere. Think again of Dante and the various tourist guides to Hades.

We can read the major religious texts from this perspective. These stories over time and through innumerable revisions have successfully come to offer truths to believers. Given the phenomenal number of believers in the world an obvious question is, can the brain do other than it does? Is this the only pleasingly untroubled way it can act given its nature and how it operates? Is this the only manner in which it can *brainsoothe* efficiently over time? Does it need, like a leg needs to walk, to act in this way to maintain its role in a species more potentially fractious than any other primate?

IF LIFE ON EARTH IS BUT A MOMENT IN THE INFINITY OF EXISTENCE, WHY GET SO EXCITED?

Memory is critical to beliefs. First, the brain has its limitations and constraints. There are limits to the capacity of its short-term memory. It's more a cash register than a computer. Some people remember much of what they experience, others very little. No one remembers everything. Even if something is remembered for a moment or two, the unhappy fact is that for virtually all people memory loss is drastic. Try as hard as you can to recall the complete details of your life at this moment a week ago, even of yesterday, let alone five years ago. New memories weaken during the day and require consolidation during sleep. (It turns out that this is an important function of REM sleep.) And when memories are reactivated following consolidation, they must be restored or kind of updated in order to persist.[24] It's during reactivation that they often are revised, when in fact memory fails us.[25] And, of course, repetition improves memory. After several attempts one can accurately recall one's telephone number without further practice. It is no longer memory but now ritual. This is surely why the same religious beliefs and ceremonies are restated so often, over and over. Even people with poor memories are likely to remember them, and anyway, few bystanders are likely to notice or much care.

Multiple squishy brain sites are involved in all of this. The hippocampus and prefrontal cortex are key functional areas. They support the consolidation of memory and its storage. Critically, these areas are in turn tied to the amygdala. This means nothing less than the immense truth that memories will be affected by emotions. There is reasonable evidence indicating that memory improves when it is associated with highly emotional moments. This sounds right. Religious revivals with their often staggering theatricality and their inevitable effect on the amygdala are implicated here. In April 2009, the inner connection of fear with memory in the brain was morbidly proved when a foolish try at a publicity photo of *Air Force One*, followed by a fighter plane as it circled the Statue of Liberty, caused thousands of people to leave the buildings close to the site of the September 11 attacks in New York City. The response was as striking as the initiative was stupid.

There are chemical counterparts to these and similar events. For example, knowing is pleasurable. Understanding a concept is thought to result in a sense of pleasure. Completing a puzzle is a form of brain entertainment. Why else would we do crossword puzzles and Sudoko? Understanding produces activation of areas likely rich in opioid receptors and the dopamine system. The brain responds to sweet ideas as the taste buds do to sweet whipped cream or genuine maple syrup embraced by a fine buckwheat pancake.

One outcome of understanding is a reduction in stress and a feeling of removing a heavy weight off one's back after an uphill hike. A kind of cognitive liberation sets in. Ideas can displease, but also please.

Yet it's not all this simple. For example, fMRI studies show that two people viewing the same information (such as a movie) demonstrate similar fMRI brain activity profiles during the actual viewing.[26] They see the same car chase in the same way. However, their personal profiles diverge significantly when they explain and interpret what they saw, much like discussions with a quirky friend with whom one has attended a movie.

As we have noted, another factor is that males and females perform the same tasks with different parts of the brain, and they perform different tasks with the same parts of the brain. The amygdalas of women fire differently than those of men. There are many factors that underlie the strong emphasis on repeated interpretation of texts and dogmas in religions. Differences between males and females certainly loom large and skilled religions have to embrace them. It is obviously necessary to develop similar interpretations among the flock even though they may all begin with different points of entry.

There are likely to be divergent interpretations of the same stimulus.[27] Religions are not just movies, although it is no accident they are so theatrically visual and intrinsically dramatic. After all, did Adam and Eve really need nudity, the snake, the apple, and all that folderol to make the drastic point about sin?

Religions are schemes for believing—deadly serious systems for those who believe. Now an interesting dichotomy emerges. Believers in the same higher authority may differ, often substantially, for example, on the attributes of their minister, how he or she behaves, and so on, much like husband and wife disagreeing about a social event. But that is only about the style of Pastor Pringle or Imam Hajdu, not about underlying reality. Disagreement about such reality will challenge agreements about imaginings.

If large numbers of group members disagree about how such imaginings as the status of the Holy Trinity are interpreted, then there may well be a formal schism, such as during the Reformation or in the separation of Sunni and Shia Muslims or perhaps in the future of the Anglican Church.

Such disagreements are intensely serious, perhaps violently so. Violence about the imagination—a quite spectacular spectacle.

Chapter 10

AND WHAT'S YOUR
BRAINSOOTHE SCORE?

Do people vary in their capacity to engage in successful *brainsoothing*? Do some people enjoy the *brainsoothe* equivalent of a spring in their step and a smile on their face? Are there luckless people doomed to depend on a sluggish cortical apparatus, which generates meager soothing relief?

Can we expect that individuals possess a *brainsoothe* score in the same way countless people assume they are somehow affected by their astrological sign? Everyone knows someone who is willing to ask a new potential partner, "What's your sign?" They may actually make significant decisions about their lives based on putative influences of planetary and chronological coincidence. It is bewildering to many astrological agnostics that astrological sections remain important features of many newspapers and other media and that astrologers continue to play significant roles in the lives of otherwise formidable players in the real world. And it is fascinating that the raw materials of astrology are events in the starry universe, which could not be physically farther away from the location of human life.

And yet there is a reality here too. Presumably ordinary people cannot abide citizenship in a random universe. So they ally with another more apparently ordered one, however arbitrary, remote, and conjectural it may be.

To such folks, would a *brainsoothe score* provide perhaps useful information about a possible course of action or choice of companion? Maybe. It's worth a try, yes? So herewith a sketch of what might be a more robust and tangible way to consider these matters. This compact chapterette is roughly but not entirely casually based on what we know about brain activity, nature, and human nature.

*   *   *

There is a story once heard in southern Oregon about Jim L. Jim and his wife had two boys when Mrs. L. again became pregnant. The pregnancy went well, but due to unforeseen complications during labor, Mrs. L. died shortly after delivering identical twin boys. It fell to Jim to name his children.

While waiting for the delivery, Jim read an article on identical twins that stressed the sameness of their genes and personalities. At a loss about how to name his boys and wishing not to favor one over the other, he named both William.

Time passed. The boys grew up. To resolve the name confusion, they became known as "1" and "2" to family, friends, and townsfolk alike. But confusion persisted. In both looks and personality they were literally indistinguishable. Often 1 was taken for 2 and 2 for 1. And in contrast to their older brothers, who were emotional, impulsive, and largely directionless, 1 and 2 were calm, reserved, and exceptionally competent. Both became doctors.

*   *   *

Whether true or not, the story is relevant to assessing your *brainsoothe* score: part of the score is in your genes.

Identical twins brought up in the same environment are very much more the same than they are different, compared with children who are not identical twins. Should they be separated at birth and raised in different social, physical, and economic environments, they are still similar but somewhat less so.

And so, at least two factors will affect an individual's *brainsoothe* score. One is an individual's genetic makeup.[1] Some infants are calm and cooperative at birth and appear to adapt to the conditions of their lives with relative ease. For other infants the story is different. From the moment of their birth, adapting to life is difficult and a challenge. For these infants, very little is easy.

Of course, there are other very important differences to consider. Some people turn out tall and some short. The bell curve of variation introduces some element of real difference among members of any population. It's a basic pattern in nature and it's where we must start in confronting the age-old issue of nature versus nurture. And then, as we have already reported, there are striking sex differences in response to stress, which may in turn generate differences in other seemingly unrelated spheres of life.[2]

Of course, genetics is only a start and not in any way final or determinative. A second factor is the environment of childhood.[3] A child who is calm and affable at birth may lose some of her calmness and cooperativeness if her caretakers are insensitive and unresponsive to her needs. And a child who at birth may be irritable and poorly adapted may gradually trade some of his jumpiness for calmness if his parents respond warmly to his needs.

Here we are describing a person's default vulnerability to stress—the platform on which rests their *brainsoothe* score, which results from an interaction of these factors. For most individuals, twins or not, the level of their vulnerability to stressful environments is largely in place by late adolescence or early adulthood. People carry this capacity to their environments. It seems almost a fact of nature that at one extreme there are individuals who are vulnerable to even minor stressful conditions. At the other extreme are individuals seemingly magically adept at living in even highly stressful environments. Like in tuberculosis, the most important determinant of responsivity to stressful environments has to do with what are called "host factors."

Brainsoothe scores, such as they are, only partially predict the aversive state of brain/body interplay. The degree of stress that exists in one's social, physical, and economic environment is equally critical. Working in highly demanding and at times dangerous environments, such as police work, firefighting, soldiering during battle, air-traffic control, or highly competitive industries (such as currency trading), invites aversive brain-body states. On the other hand, those working as librarians, secretaries in large corporations, or as window washers for one-story buildings usually experience only modest stressful moments in their work.

As well as what they bring to the party, people live in multiple environments with different levels of stress. Home may be minimally stressful but the workplace highly so. It may be the reverse. The possibilities are endless. Moreover, the degree of stress may change in the same environment over time. Couples in a troubled and stressful marriage may resolve their differences and find that the stress associated with their relationship has largely dissipated. Or someone is gracefully employed in a minimally stressful job but then suddenly is thrust out of work when a recession strikes. The environment has changed from soft rose to angry purple.

Then there is the capacity to *brainsoothe*. This capacity seems largely independent of *brainsoothe* scores, whatever these might be. There are individuals who are highly vulnerable to aversive environments yet also highly capable of *brainsoothing*. But other similarly vulnerable individuals are poor soothers. Myriad explanations have been offered for these differences—Freudian, Jungian, cognitive limitations, intelligence, training, and so forth—but compelling data are few in number. Those findings that are available point to an interaction between genes and upbringing, which is hardly a surprise.

FINDING YOUR *BRAINSOOTHE* SCORE

Is there a way to identify your *brainsoothe* score even though your local newspaper hasn't a section providing you daily recipes for how you should spend your day based on that score? There is. And it is totally arbitrary, conjectural, and questionable. Yet it may also offer a practical and user-friendly way to evaluate what appears to be an important factor in how people conduct their lives. Let's have some fun.

Using Scale A, give yourself a score for the average number of times each day over the past year when you found yourself frustrated, angry, and/or thwarted when you were in minimally stressful environments.

Scale A: Times per day that you are frustrated, angry, and/or thwarted in minimally stressful environments

1–2 per day	3–5 per day	6–9 per day	10 or more
☺ ☺	☺ ☺	☹ ☺	☹ ☹

The frequency of your anger and frustration combined with the number of times you are thwarted when the environment is minimally stressful is a convenient proxy for your *brainsoothe* score. (Of course this is made up, as is this entire chapterette.) A score of 1–2 per day is low, which means that you are minimally vulnerable to the normal and random stresses of everyday living. With a score of 3–5 per day, you are close to average but still a bit below. This means that you are not affected by many of the normal and unavoidable stresses of everyday living. But a degree of vulnerability remains nonetheless.

With a score of 6–9 per day, you are haplessly above average. Even in minimally stressful environments you are vulnerable and often likely to experience aversive brain-body states. With a score of 10 or more, you are at the high end of the vulnerability scale. For individuals with this score, aversive brain-body states are the rule, seldom the exception. Again, we are being arbitrary here. But kindly bear in mind that we begin with real conditions and histories, not the happenstance locations of Pluto or the moon.

Then, of course, there are also the *real* environments in which people live, and these can range from minimally to highly stressful. Using Scale B, give yourself an average score on the stress level when all of your environments are added together.

Scale B: Average stress level of your environments

	Low	Moderate	High	Intense
Stress level =	💣	💣	💣	💣

If your score from Scale A is ☹ ☹, in an intensely stressful environment (💣) you will need optimal capacities to *brainsoothe*. On the other hand, if your score is ☺ ☺ in a low-aversive environment (💣), average *brainsoothing* capacities may be sufficient. Or if your score is ☺ ☺, in any environment (💣, 💣, 💣, or 💣) it is likely that you will successfully *brain-*

soothe—deliver that spring in the step, and so on. And, we stress, most individuals come to know their *brainsoothe* score, the stress intensity of their environments, and their *brainsoothe* capacity.

Is this crazed or cool or what? What is to be made of it? Is it akin to astrology or ideological politics and thus another package of overcomplicated speculations for which there is little or no compelling evidence? Perhaps. Yet this analysis is consistent with experience and with massive amounts of psychological and psychiatric observations. It is clearly established scientifically that there are significant individual differences among people in their vulnerability to stressful environments and their capacity to withstand them with what we have called *brainsoothing*. And there are, in fact, hospitals for persons unable to cope, at least temporarily, as well as agreeable meeting places for those who cannot only endure other people in a soothing manner but also seek them out.

Where is religion in all of this? Do individuals who are highly vulnerable to stressful environments flock to religion? Is there a relationship between the level of stress in an environment, such as in post-Katrina New Orleans, and the likelihood that a person living in that environment will seek religion to facilitate *brainsoothing*?

At this moment it is, of course, impossible to answer such questions or virtually any other of such magnitude. But possible answers are suggested by a number of facts. As we've already noted, at least 80 percent of the adult world professes religious affiliation and a large proportion of these people actually engage in observable religious behavior, such as praying, attending temples, wearing insignia, claiming affiliation, and contemplating some versions of heaven and hell.

Who would deny, after all, that daily life is episodically, if not chronically, potentially stressful? The number of people who report they are satisfied with and comforted by their reli-

gion exceeds those who are dissatisfied by a ratio of twenty-to-
one (though self-reports may be very unreliable). Terminally ill
people rarely forsake their religion. Those who have lived
without religion throughout their lives embrace a religion
during their final days often enough for it to be interesting.
Deathbed conversions are not merely vaudeville comedy. As
the thin string runs out, it is gripped ever more tightly. And
those who have lived through frightening and life-threatening
experiences nearly always attribute their survival to the intervention of a higher power and frequently thereafter embrace
religion.

Even the remarkably talented and remarkably raucous
rock guitarist Eric Clapton offers in his autobiography of 2007
a vignette about a time of great and strenuous uncertainty in
his life, in which he is visited with an experience he finds no
other way to describe than how he defines divine. For decades
he had suffered through extensive bouts of alcohol and drug
addictions. It was clear his cortical function was afflicted by
external factors beyond his individual control. Eventually he
succeeded through a therapeutic community to achieve
sobriety. He decided to go sober and permit his personal inte-
rior brain to benefit from its own natural gifts and not the con-
tents of bottles and envelopes.

He decided this choice arose from forces beyond his own.
Dramatically exceptional though he is as a musician, as a
searcher after *brainsoothing* his remedy was not exceptional
but par for the turbulent course of the lives of many.

Our effort to confect a quasi-numerical value for *brainsoothing* has been a bit of a sport but not merely a pastime.
The utterly phenomenal strangeness of the fact that people
build huge cathedrals or tiny chapels and attend and share
them and learn how to behave in and outside of them suggests
an equivalently potent need is being satisfied or at least
addressed. The brain may in the beginning be a lonely hunter

for soothing, but in the end it returns to the fold—to the realm of civility, company, familiar other people hanging around, a bit of music and swanky costumes, old books of looming authority, an annual as well as daily program. Something, almost anything.

We hope we have been careful enough not to oversell our assertion that the power of religion cannot or should not be attributed only to the idea of an all-powerful deity, *the best evidence of whose power is what we subordinate humans ourselves do, say, and believe.*

But we cannot emphasize enough that the explanation for this immensity of human action is not in the stars or movements of the moon but in our hearts, and more specifically and exquisitely, our brains.

Chapter 11

RATHER A BEGINNING, NOT A CONCLUSION

T he world of religion is an intricate, colorful jungle. In what you have just read we have tried to find paths into and through it, and we have also sought as fairly and incisively as we could to identify what is happening to the human person at the center of the story.

Of course, we have focused on religion because that is our specific subject. But the overall subject is the jungle itself. The persistent fact is that virtually everywhere we look, in environments that are utterly different and rest on social histories of incomparable variety and densely different trajectories, nonetheless we continue to find recognizable versions of the central religious theme. Endlessly and sharply, variable societies reiterate an ancient, recurrent theme. Like the normal temperature all nationalities read on their thermometers when they aren't ill, it's a predictable commonality. They devise and sustain ways to *brainsoothe*. Nearly always religion is involved.

So we have been compelled to try to isolate what factors might or must be common. Both because of what we know as professional workers, but more because we think it's correct, we have focused on the brain as the organic manager of the behaviors and groups we all know as religious. Some 80 percent at least of the world's population is in one way or another roiled, turmoiled, and apparently soothed by some form of religion. At least they are sufficiently *brainsoothed* to keep it in the inventory. It seems worth the fuss, the tithes, the mosques, the diets of Lent and Ramadan, the yielding up of almighty autonomy, the fear of hell, and the difficult lure of good behavior.

This is an astonishing fact. Almost by definition the basis of the behavior is explicitly and defiantly supernatural. It is literally out of this world. We have to assume this is largely by preference, which for many is a great virtue and for many an undeniable provocation as well.

Within the protective, airy cylinder that surrounds the phenomenon—an ozonelike layer of piety—there appear to be at least 4,200 known religions and probably more. They strongly influence the inner and outer lives of those committed to them. They have direct effect on how societies are organized and governed. We can look at them in perspective. It will appear that a magnetic force operating from some distance away exerts pull and forces push on the behavior of believers.

There is an unseen choreographer in whose ballet they are happy to dance. They are also willing and able to experience emotions and sustain thoughts that are presumably animated by the religious magnet. Finally, this is amazing. And when they engage in the many rituals that compose the multitude of religions, the idea of choreography ceases to be a metaphor. It becomes what is actually happening. The dance is life itself, however many years it may be.

And, of course, religions have an impact on active nonbelievers. If nothing else, the calendar of holidays and observances of one group—think of Christmas—will have an effect on the business and pleasure of people coolly immune or even hostile to what the religion is all about. The imposition of temperance about alcohol by fundamentalist Christians and nearly all Muslims is a clear case of control by a religious theory of even private behavior of people in their own dwellings. Again, a clear display of the impact of believers on others is in Jerusalem, where the Muslim holy day is Friday, Jews observe Saturday, and Christians, Sunday. The city takes a deep breath nearly half the week. The pace and focus of urban life in that iconically storied city is delineated by what believers do and believe and, therefore, how others must respond. Often the response of nonbelievers to the demands of religion is courteously tolerant and vice versa. Then it may also become legally draconian and, as during the Spanish Inquisition, perilously dangerous.

Nevertheless, in countless communities there are persons who are uncertain about religion, skeptical about it, or angry with its impact on their lives. Or they may be atmospherically fearful or trepidatious about the mysterious forces of the universe about which they have been warned by content and pious neighbors who enjoy the certainty of belief. Whatever their sense of the plausibility of religion around them, they have no choice but to adapt to the reality of what they may perceive as other peoples' choice of what is useful.

Presumably, laws or consensual practices establishing relations between church and state are ways of managing this issue. This was certainly the intent when the United States Constitution was framed for a group of colonies comprised of avid members of an array of religious traditions brought from Europe. For the United States, religious and nonreligious ideologies have changed, and religions have not been immune from their diluting effects.[1]

The more telling point, however, is found in places where the power of the state and of religion are merged into one entity, such as in Saudi Arabia. Here the politics may be simpler, but there can be genuine danger for persons who wish to segregate themselves from the dominant majority. Hence, individuals in Saudi Arabia may be executed for nothing more than simply owning a Christian Bible or for converting from Islam to another faith. The exertions of the Taliban may appear extreme to outsiders, but they possess implacable logical correctness within their system.

Related to this are the less dramatic but still controversial efforts, for example, of some American polygamists, who defend their sexual practices with the doctrine of religious freedom for a branch of Mormonism. Or Muslims in Canada who have claimed with occasional success that they should be governed by the foreign traditional laws of Sharia and not those of the country in which they live and, as a matter of fact, to which most Muslims have voluntarily moved.

It appears to be almost conventional that some groups are so convinced of the accuracy and importance of their religious beliefs that they are prepared to sanction sometimes violently those who don't share those beliefs. The rationalist assumption of many freethinkers of the nineteenth century was that religious certainty was becoming less and less effective as a governor of action. Surely some formula such as utilitarian secularism would become the dominant choice of sensible societies. In 2009 a citation from Bret Stephens of the *Wall Street Journal* noted that the founder of Pakistan, Muhammed Ali Jinnah, predicted in 1947 that "in course of time Hindus would cease to be Hindus and Muslims would cease to be Muslims, not in the religious sense, because this is the personal faith of each individual, but in the political sense as citizens of the state."[2]

Of course, this hasn't happened, and not in Pakistan alone. In fact, there have emerged epically broad, even worldwide, lethal swirls of conflict founded on religious difference. These are often catalyzed by the inevitable link between religion and the ethnic regions within which different ones prosper. For example, the Harvard professor of comparative literature

Ruth Wisse has argued in an extraordinary book that anti-Semitism—of course a religious impulse attached to a tiny, almost nonexistent group of people—has been the most durable and entrenched ideology of the last thousand years, more continuously effective than others, such as fascism and communism.[3] More recently, it mobilized nothing less than the great German imperial apparatus to a disastrous convulsion and now preoccupies vast Islamic populations, many schools within which youngsters are taught active contempt for Jews, especially those living in Israel.

RELIGIOUS HATRED IS HARD WORK AND IT IS OFTEN EXPENSIVE TO EXPRESS IT IMPRESSIVELY

So why does it exist?

In chapter 9 we discussed the puzzling nature of many of the issues we raised about religion, and here we encounter another: Why? Why bother?

People who could otherwise go to a beach or buy shoes or sleep or do Sudoko or spray chemicals on roses decide it is vital and desirable that they punish their fellow inhabitants and, if necessary, attack and kill them. They may even cross boundaries and oceans to accomplish this. They appear to find it very worthwhile and even satisfying.

But killing people without benefit of modern military hardware or poisons is difficult, and time-consuming. It may well involve dangerous and messily bloody personal encounters. It also can absorb funds, which could otherwise be used to go to a war movie, buy two kinds of goat cheese, or purchase a biography of Elizabeth Taylor.

Even with fine contemporary military gear, warfare often involves considerable uncomfortable travel to completely uninviting places. The warrior will on triumphant or reluctant arrival have to begin to endure living conditions that would be illegal were the cause not thought to be so important and valuable. It may become impossible to maintain agreeable contact with the people one loves and craves. And, of course, since virtually all people are deeply unwilling to be wounded or killed, the infidel enemy will fight back fiercely. This becomes squalid and dangerous all around.

And for what?

Many traditional economists are fond of telling us confidently, if wistfully, that human beings operate rationally. Economies are thought to be founded on the fact that people make sensible choices between one product or career and another, one dwelling and another, and one fax machine model and another. Overall, they are supposed to live their lives in a manner that responds to general rules of day-to-day rationality. But what about the young jihadi who leaves his family to go to an isolated training camp in Pakistan to better learn why and how to destroy Sunnis or westerners or Hindus? What about the Muslim neurologist who elects to drive a car of explosives, as one did, into central London to accomplish a great purpose and audibly invokes Allah as flames surround his body? What do eager martyrs of any religion think as they fully perceive the immensity of their choice at the perfect point when they succeed in their mission?

And this is not solely the situation of religious believers of now and then. There are countless secular exemplars of the important notion that so long as a nonbeliever in the revolution or modernization or precise progress or return to the past continues to draw breath, all is lost. For example, in his rather riveting novel *The Gods Will Have Blood*, first published in 1912 about an exuberantly lethal proponent of the French Revolution, Anatole France describes his adaptation of the postrevolutionary slogan engraved on the Palais de Justice in Paris: "Liberty, Equality, Fraternity—or Death."[4] The "or death" formula has been a popular favorite in countless social movements designed to improve human matters. From the gulags of the USSR to the Spanish Inquisition, the political ministrations of Pol Pot, the Chinese Cultural Revolution, Fidelismo, and every which other way, there has long been support for the validating and cleansing utility of killing real or putative enemies of the great new day. Wonderfully, blood turns to elixir. The genocidal Nazis were not original, just better organized, more efficient, and more industrial.

Dare we raise the possibility that this is the most extreme version of *brainsoothe*? Things get so bad that only worse events can palliate them? How could this be?

Life is marked by perturbation. The dream of smooth perfection is endlessly sullied by the inevitable events of life, such as trying to become educated, raising children, responding to aging parents, undergoing periodontal surgery, and enduring economic and political conditions that cause irritation and perplexity.

Why don't other people just understand and accept the right way to do things and the right way to live? A celebrated brace of confident holy texts have advocated this with great assurance? But still people don't. Why don't they accept the religious authority of traditional leaders and do what they say and accept? As Ruel Marc Gerecht has stated, "The so-called June 12th (of 2009) revolution is the Iranian answer to the recurring hope in Islamic history that the world can be reborn closer to the Prophet Muhammed's virtuous community."[5]

In this book we've emphasized that for some reason the brain is poorly designed to deal equably with prickly realities —perhaps because it evolved in a simpler world than the one we inhabit. It appears to prefer the perfection it has created in the guise of heaven and in various secular utopias or even championship seasons of sports teams.[6] These wonderful accomplishments will surely banish irritating features of life and lure the smooth ones to flow endlessly to the fortunate participant. Utopia becomes I-topia. However, as Iranians and others have found, they do not.

A signal of how muscle-bound the brain may become when faced with perturbing realities is that fighting religious wars and embracing martyrdom may episodically appear to provide solutions to the ills of life and of a person's life. A flailing and abrasive life in the moment is the flip side of the dream of heaven. Current life may be felt to be so miserable that the only antidote is self-destruction along the well-marked route to that perfect scheme of heaven.

This is not common suicide, that act born of personal despair and surrender, not theoretical assertion. No. Martyrdom is a transition to an elegant, perfect, comforting parlor. And recall briskly that this heavenly perfection has been evidenced by no one except those who want to go there, expect to go there, or have made it their life's business to explain to other people that with appropriate and faithful action they can get there too.

It's nothing less than remarkable. We keep coming back to that word. It's about as unexpected a phenomenon as the Metropolitan Opera with all the countless people, events, and artifacts that compose it. Should someone ask what incredible stimuli rise up at the corner of Broadway and Sixty-Sixth Street in Manhattan, the answer could as plausibly be heaven as the world of opera.

THINKING OF HEAVEN

A theme of this book is that the brain perceives many things. It conceives of higher powers, superior forces in authority, and unseen, perhaps fugitive but finally important influences. It appears also to have the capacity to attribute truth to its perceptions. Like a notary, the brain certifies. In countless cases it can believe symbolic phenomena with the same intensity as real and palpable facts and events.[7] The brain secretes belief. As we've indicated, such belief is often the basis for actions. These may be intimate, as in personal observances of piety or commitment to monogamy, or massively social, as in the society of the Catholic Church or in hazy belligerent reveries about the restoration of the Muslim Caliphate.

Belief, not doubt, is the brain's default.

There are studies that report experimentally induced religious thoughts are associated with reduced cheating and increased altruism toward strangers.[8] But what takes place in the laboratory is not necessarily repeated in the street, where the data are more consistent with what we have suggested: namely, that once a belief is established as acceptable, it may be very difficult for people to do much about it. For example, if it is decided there is life on other planets, then that becomes the operating theory with which people function, only a tiny number at a moment. Similarly, if it is determined that there is a heaven, access to which depends on appropriate earthly behavior, which is mediated by religious figures and then divinely judged, then that becomes the guide to what people who are committed to the system do. And there are many, many such people.

This is something about which they can take action. It is something about which they have efficacy. They can buy indulgences. They can forego a secret love affair in another country, fast on holidays, or chant some special words. The quest for religious merit can occupy almost as many resources as they wish to dispense. It can use up as many hours of the week as they can spend and enjoy and justify.

The skillful brain of our species can clearly recognize as well as confect a variety of fears. It can confront an array of ambiguities and sort them into a manageable scheme.

We've outlined some of the brain mechanisms that derive

from the experience of uncertainty and of facing the unknown. These are features of daily life everyone knows. We have outlined how they can lead to chemical changes in the brain, in its secretions, and in how they interact with other bodily processes. They are common sources of stress that produce aversive physical and psychological states. In turn, the crux of our argument is that people commonly seek to defend against such aversive states. They do so by trying to *brainsoothe*.

Many of the details of the scientific findings that we have cited will undergo revision, some on a weekly basis. That is the healthy, imperative way of science. But this should not be confused with our central message about the core of the process: it's the brain . . .

An important way of doing all this brainwork of managing anxiety and outright fear is through religion, which is a creature of the brain's system of belief. Please note we make this comment with neither censure nor approbation. It is simply what we think is the reality in this realm, a reality with remarkable ubiquity and persistence and importance.

We stress again the critical significance of a nonjudgmental perspective on these matters. We want to emphasize what we have intended: a balanced approach to the matter of religion. It has for too long stimulated remarkable bouts, pro and con, about its nature, meaning, and value. But it has produced neither a fight card nor a dance card that interests us.

ONE, TWO, THREE: THE SECULAR TRINITY

Let's review briefly how religion may serve the purpose of calming an easily perturbable human being and the group in which that person lives.

We've identified three principal means: positive socialization, rituals, and beliefs.

The affable, unthreatening, repeated, and easily understandable social learning and familiarity associated with religions induce the biochemical system to abandon ominous
fears. Biochemistry becomes a censor. And it appears to induce
nonstressful profiles of desirable outcomes. There seem to be
abundantly clear neurochemical benefits from enjoying the
menu of positive socialization. There are social interactions
that are available to religious persons in their devotional
groups—interactions in preparation for which they may have
had hours and days of training and repetition.

For their part, rituals palpably relax the body. They generate a protective bubble, a form of behavioral spa. They
reduce the concentration of stress-linked chemicals, such as
cortisol, in the body. Among other effects they abet reduction
in blood pressure. Any activity that produces such an agreeable impact on the body will induce the body's owner and
master to seek out the conditions that will produce it. A powerful suggestion about the naturalness or the physiological
depth of religion is the fact that ebbing blood pressure is also
associated with immersion in natural vegetation and very
clearly with caressing a pet animal. These effects begin to
explain the remarkable prevalence of house plants and pets
living in dense urban environments. It is presumably significant that gardening is the most popular of North American
recreations. And the population of pets is amazing. Stroll the
aisles of a supermarket and you will see that there is almost as
much shelf space for pet foods as for human biscuits.

Does blood-pressure response to pets and plants tell us
something about naturalness? And something about the naturalness of religion too? And isn't it the likely basis for the
strong and, in many cases, research-supported presumption
that religiosity equates with health and increased longevity?[9]

Finally, in the context of higher brain functions, religious
beliefs appear to succeed in simplifying the complexity of exis-

tence and social life. They carry a knack for reducing the ambiguity and the uncertainty surrounding this complexity. Whatever the factual status of the sources of clarity and cognitive order, nevertheless, there is evidently recurring satisfaction in encountering over and over sacralized explanations and descriptions of life's buzz and fuss. Belief appears to produce for the brain a comfort zone similar to what ritual generates for the body.

THE EXPLANATION PUZZLE

An animating reason for writing this book was our discontent with the most salient explanations of religion's power and incidence. It seemed vital as well as interesting to explore the matter. Our notion was that, so far, there was no wholly satisfactory explanation. While partial stories had been very earnestly told, we were not satisfied there was a clear and embracing narrative that we could follow.

But why? We concluded people had been looking in the wrong places. For example, one claim was that religion had evolved as a biological fact because it had improved the opportunities for religious groups to survive and prosper while the irreligious floundered. Another claim was that a version of genetic group selection was thought to operate and that religious behavior served the effective functioning of groups.[10]

These were partially attractive accounts of what happened. But given the clear link we think we have stressed very appropriately between brain function and religion, it seemed more efficient and economical to look at what religion does for the brain. How does it serve what we've called God's brain? And how does the idea of god serve the human brain?

Again, insofar as the brain is both the source of religion and the principal recipient and manager of its stimulus, then surely the brain had to be the explanatory lever that generated understanding of why it existed and what it did. And in our analysis of what the various elements of the brain did and how the neurotransmitters and other substances operated, we were confident a better explanatory mousetrap was at hand.

One limited but historically interesting and useful explanation, which provided some insight into what was involved, was existentialism. In essence, this was a suggestion that some scheme of explanation and belief could reduce the uncertainty and incipient confusion and defeatism that attended the facts of life and living, but with no overriding religious justification.

This was certainly an understandable intellectual response to the despair of World War II and its aftermath. Introducing sense into senselessness was an attractive act of assertion as the stones of destruction began to be reconstructed.

But existentialism generated no durable and appreciable social institutions, no interesting music, no inviting architecture, nor any furnishings comparable to effective religions. And as a response to the local, if vast, disaster of World War II, it retained a Euro-American focus in a relatively small social milieu during its limited day in the sun and the dusk.

Perhaps a stray graduate of University College London or the Sorbonne might return to India or Lima or Kyoto to represent the existentialist core. However, the European origin of the body of ideas was an inevitable burden on its translation to communities with very different points of philosophical departure. What has reincarnation to do with existentialism?

A very different focus of attention to causality seemed essential. So as one result of a revolution in empirical primatology, along with others,[11] we were able to propose that both humans and chimpanzees shared a common scaffolding for religionlike behavior. Similar behavior, brain structures, and comparable patterns of neurosecretion have been identified for both species. Suddenly we could identify a new, broader, and deeper form of analysis of the mysteries of spirit. And in a glad spirit of a potentially liberating synthesis, we proposed that major elements of the potential for *brainsoothing* were evident among both species. We were different but not alone in the world.

The differences are of course huge. One colorful and extensive one is that humans have generated an immense inventory of beliefs and theories about the meaning of existence. It's possible that so do chimpanzees. But we've no way of knowing if they do or not. They don't write things down or stain glass or sculpt saints or wear colorful garments on Sunday morning.

A mighty skepticism about their formal religious verve is surely in order. Chimpanzees seem to display no acknowledgment of an identifiable and potent deity/being that directly affects their behavior. They display no moments of what might be seen as connection with more powerful forces that operate in their day-to-day world. Even gesturally, they do not appear to make the kinds of signs of deference to higher symbolic authority, such as raised hands, hands folded in prayer, faces turned heavenward, or other universal forms of human obeisance, such as bowing. However, there are reports of female chimpanzees clapping their hands to assuage moments of group turmoil. So far as trained human observers can tell, they do not invest particular places, trees, rocks, rivers, and so on with unusual significance. They don't make and manipulate objects that command special deference, such as altars, icons, and crucifixes.

In contrast, the human knack at formulating and obeying higher authorities is an immeasurable step away from whatever explanatory theory of social order chimpanzees may share among themselves. So far as we know, chimpanzees keep such reverential notions stubbornly private as they navigate their groups and their days and nights. This is not to say they fail to possess the kind of moral sense or code—they certainly do— that is often thought to depend on religious values.

Primatologists learn more and more about prosocial behavior of other primates. There have been countless observations of group-sustaining chimp behavior, of cooperation, of sharing, of acknowledgment of loyal group membership, and the like. One of the more colorful and potent controversies in the sciences of animal behavior has been the often-successful effort to demonstrate that animals possess a sense of fairness, of sharing, and even of individual dignity. We discussed at some length in chapter 5 ("Religion as Law and the Denial of Biology") and proposed a new identity between science and morality expressed in a burgeoning literature. When we recall that the earliest universities were essentially established for theological purposes and to produce the leaders and managers of churches, who educated their followers, the change from then to now in the interaction of knowledge and belief is striking.

We have to consider carefully, indeed, the claim by religious leaders that obedience to a higher authority is the sine qua non of moral interaction; that without an operating god, all is recklessness and even whimsy.

Or, paradoxically, are we the lawless, fallen ape if we need an all-powerful referee to cause us to act with decency, whereas humbler chimpanzees manage convivial social interaction on their own hook, with not a single chimpanzee in uniform, reversed collar, or judge's robes?

Again, in possible contrast to chimpanzees, or, for that matter, any other species that may have some capacity to formulate norms of the divine and the supernatural, human beings have a genuine skill at creating higher authorities and attributing to them sources of truth and authority. Countless communities have built countless places of worship and centers of deference to make the impulse of their beliefs material and real. Even nonbelieving tourists frequently find them impressive and rewarding.

The human brain is good at this, excellent in fact. It can

generously produce expectations of how people should behave and why.[12] Bolstering these expectations is a well-advertised catalogue of the imminent and even inevitable current and future consequences if particular mandated behavior is not forthcoming. In symbolism and, for many, in practice, the summary of a person's inner thoughts, feelings, and behavior is a ticket to ride stamped either Green Heaven, Red Hell, or Yellow Purgatory.

Not only is behavior required, but states of being, too. One of the most challenging obligations of those committed to a religion is that they experience states of grace, purity, loyalty, charity—a long list of words. Often, the more important a particular religious word is, the more imprecise and global it is. But the final impact, the last destination entry on the itinerary of the ticket to ride, is what a person is, not only what he or she does. One's state of being is a telling feature of religious destiny. This is a challenging requirement. The potential sinner has to be the ongoing performer and judge of the matter. Like a tightrope walker, he has to keep balance, ignore the crowd, look ahead, and not look down—ever.

We have already described the swift surgical utility and effectiveness of the concept of guilt in religious operation. It provides a clear guiding metric, for example, to Catholics, Jews, and Muslims, who are provided an inventory of good things to do and bad ones to avoid. Among Jews a *shonda* is a shame and a sin, while a *mitzvah* is decisively pious and valuable. Other religions have very similar ways of accounting for decency.

It seems virtually certain that guilt as a factor in emotional life became significantly more important after the Protestant Reformation. Of course, the individual soul had more moral autonomy than before. But it also faced correspondingly greater peril. This was a special plight for Protestants who no longer had readily available the smoothly exculpatory technique of official confession. With confession, whatever amends had to be made were on a private ledger. The confessional booth could hold only one soul. The record was created and audited and in an important sense usually ratified by the same figure of authority, however frail he was.

But now there wasn't even such an accountant to turn to. Only the Protestant self now counted. Stand or fall. Do the right thing or the deficit is clearly and crisply yours. This was an enormous burden, come to think of it. But according to historians of religious sociology, such as Max Weber and Ernest Gellner,[13] it also offered to Protestants a capacity for muscular self-interest. In turn, the proposition has been made that this was intrinsic to the emergence of the spirit of capitalism.

This is a very large argument, indeed. Nothing is clear-cut. However, for example, the differences in economic performance between French Huguenots and French Catholics, for example, and French Canadians and English Canadians suggest that moral freedom was more than ever useful to bankers and other entrepreneurs rather than priests in the new post-Reformation society. The balance sheet of life had to merge into a shifting balance sheet of business profit or loss. Here was new, worthy weight to celebrate as it perched on the scale of virtue—a weight lightened when the soul involved could expect that heavenly reward was firmly signaled by worldly success. In a sense, Calvinism offered a generalized confessional.

Doing well implied being good.

The clattering role of dynamic capitalism aside (even though it changed the world), it is small wonder that the small-town courtroom image of the ur-arbiter Saint Peter at the pearly gates is so popular and ubiquitous. It is intriguing and finally humbling that humans have been able, when we've bothered, to imagine the lavish architecture and experience of the pearly gates, but we've apparently been unable to confect an accurate accounting system with comparable grandeur. However, this should not be surprising if we acknowledge that everybody in this system is a vulnerable and humble shop-keeper of his or her own moral inventory.

And for nearly everyone the list of wrongs is likely to be modest and hardly dramatic: one glance at a school chum's exam, one pittance of income concealed from the taxman, one surprisingly exuberant but evanescent flirtation, one bad word to the homely coworker with bad skin, one surely disconsolate opinion of the rookie cleric filling in while the beloved local curate is attending courses in the capital city.

Small-time stuff.

WHAT WE'VE NOT SAID

We have often enough stressed our claims about the limits and character of our approach to the scientific question, *Why religion?* We'd like to reemphasize what these limits are, why we have insisted on them, and where they stand in the larger worldwide seminar about religion and contemporary life.

And we also want to anticipate some possible responses to our argument, which we can express very simply.

Readers of this book who believe in a God can on its basis affirm that they are sensible folks pursuing a course that is robustly based in nature and connects to the supernatural. We hope that nothing we have said should demoralize or irritate or stymie persons who conclude that if the brain is the prin-cipal target of religion, surely this demonstrates that religion is a higher-order phenomenon and that its accord with contem-porary brain science is a fine asset.

On the other hand, those who are confounded by or oppose religion will find comfort in our avoidance of confident assertions about divine truth. They may choose to recall warmly Karl Marx's earlier formulation on the matter, which was that religion is the opiate of the people. We have extended the metaphor to include all people, not only the working class—in fact, often elites may be more severely restricted by religious authority, if only to support their legitimacy. And our relatively firm location of religious activity in the skull is presumably a sufficiently secular denial for them of the elaborate and ambitious claims of religious advocates.

Here are some statements we have not made.

Authority

We have not said that a higher authority exists or does not exist. We have no reliable and comfortable way of knowing the truth or falsity of assertions confidently made over centuries by countless people. We understand fully that many will see this as a central, even profound, failure of imagination, piety, skepticism, and generous conviviality. But we are stuck here. When the writer Isaac Bashevis Singer won the Nobel Prize for Literature, he was asked by a reporter for the *New York Times* whether he believed in free will. He replied, "Do I believe in Free Will? Of course. I have no choice." We've had no choice about higher authority given our training and professional work, standards of evidence, and caution about zealotry. We concede a simple inability to provide an answer to the question, *Why religion?* which at the same time provides its own answer by denying the need for the question in the first place.

Comparative Religion

We did not regard it as our business or within our competence to announce which religions are better or worse and which forms of observance, piety, and agnosticism are more or less desirable. Not only is such comparison mongering spongy and too easily self-righteous as an analytic matter. It is also, of course, the hot source of endless arguments and contests, which we have been keen to avoid. A fair amount of such contestation has yielded religious warfare so widespread, so evidently satisfying in a morbid way, and yet very dire. So we included ourselves out.

Doing *Brainsoothing* Well

But there may be one legitimate focus of comparative assessment of different religious groups. That is how successfully they manage their affairs to provide their members more or less effective *brainsoothing*. There are some obvious second-raters in this contest, such as the austere American sects who did not believe in sexual reproduction and whose membership ebbed, clearly not only because of simple demography but also because their members may have been punitively bored and missed out on a lot of the glad fun associated with sex. In effect, they failed to engage in a suite of activities including playing with babies and children, which for other groups provided engrossing reward and vivacity.

We have already noted what may be the special efficacy of the Catholic confessional as a source of *brainsoothing*. Surely, architecture, music, and costume will also have impacts on the brain. And surely also, financial aspects of membership, such as tithing, will enhance or distort *brainsoothing*. This will depend on how fair and effective they are in delivering the structures and occasions within which *brainsoothing* can occur. Does a religion provide a suite of occasions and services that supports the predictable and legitimate needs of ordinary people, and does it do so at a fair or at least sustainable cost?

Some monetary matters, such as the sale of indulgences, may in retrospect and to outsiders seem unusually exploitative. Nevertheless, for those at the time with money to burn and a burning hell to avoid, the investment may have seemed at least prudent, if not comfortable. As well, tithing—giving an ongoing percentage of one's income to the church—has, of course, been a robust mainstay of the helping systems in many communities. It is impossible not to acknowledge the role of religious groups in supplying resources and assistance during times of great need. In the United States in the aftermath of Hurricane Katrina, many observers compared most unfavorably the behavior of secular authorities alongside religious ones who served directly and right away. In the post-hurricane Burma of 2008, the contrast could not have been greater between the helpful behavior of religious monks and the bewilderingly ugly inattention to human suffering of the secular military government.

What about Believers?

We have not said that believers are crazy and suffer from delusions and mental deficits. Since, as we have observed, nearly all human societies indulge in some sort of religion and belief, this perspective would commit the entire species to insanity, a difficult position for naturalists, which, however, few have sought to sustain. At the same time, for nonbelievers, belief remains a puzzle. The great sociologist and historian of three major religions, Max Weber, described himself as "religiously unmusical." This phrase will resonate with many nonbelievers. Nevertheless, it is unlikely Weber would assert there was no music even if he did not hear it.

What about Telling People What to Do?

We have not said people should believe or should not. We have tried to show that belief may produce amiable impacts on the brain. But seeking a particular flavor or form of it is wholly an individual choice, except in those many situations where religious indoctrination begins at the earliest moments of life. For some people, the process of religion offers a broad range of social opportunities, which they might fail to enjoy otherwise. At the same time, immersion in religion may be a curse and a hazard for other people—recall poor, priest-ridden James Joyce.

We have also explicitly not said that religious belief is a necessary basis for moral and prosocial life. Decency does not emerge only from holy writ or pastoral admonition—recall Chief Fred and the members of his group. That is a claim often enough made by protagonists of religion. But it is more likely an exploitation of ambient human uncertainty about the future than an essential animator of moral communities. History shows that religion is a purveyor of moral order but not its originator.

We have failed aggressively to provide advice or opinion on how to change or not change religions or groups opposed to religion. That is part of another agenda, not our own at this time.

Accordingly, we are neither optimistic nor pessimistic about the future of belief. However, we have acknowledged the despair that surely should follow the passionate connection between belief and active and often violent bellicosity.

We have very clear notions about the respective virtues of belief and scientific certainty about such matters as constructing and flying airplanes. We have a similar regard for medical science compared with Christian Science as a treatment plan. We have been clear about the need for and value of evidence on the basis of which to conduct action and build and operate things. We remain mystified at the claim that it is the very absence of evidence of something that is the firm proof of its existence.

We have said we cannot provide evidence for the existence of the forces and experiences that are at the core of religious assertions. However, we have also sketched ways in which such assertions can and do have practical effects on the lives of people and their communities.

Again, a large company of theorists and commentators on these matters is already hard at work and we cannot and do not want to compete with their often nuanced (but occasionally nutty) craft. Consequently, we have not seen fit to provide announcements about whether religions should or should not change, and how they should do it.

We hope we have responded to the worlds of religion and belief in a useful and equable manner. But we have not—not!—taken for granted or assumed the fact of a higher authority. Could anything be more angular but important in a book on religion? Our obedience has been to the law of parsimony: to explain the facts of nature with the most basic rather than the most exalted set of causes.

If God is a creation of the brain, then God's brain is our brain. There is then no lower authority to be found than the operations and impact of our brains and the process of *brainsoothing*. We named the brain as the source of infinity. This is surely appropriate since it was our commitment to that brain that caused ambitious humans to call ourselves sapiens.

And, by and large, that we are, give or take . . .

ENDNOTES

CHAPTER 1: AND WHAT AN AMAZING, IF IMPROBABLE, STORY IT IS

1. National and World Religion Statistics, 2009, www .adherents.com (accessed June 6, 2009).

2. R. Dawkins, *The God Delusion* (Boston: Houghton Mifflin, 2006); C. Hitchens, *God Is Not Great: How Religion Poisons Everything* (New York: Twelve/Hachette Book Group USA/Warner Books, 2007); S. Harris, *The End of Faith* (New York: Norton, 2004).

3. D. Overbye, "Wisdom in a Cleric's Garb; Why Not a Lab Coat Too?" *New York Times*, June 2, 2009.

4. M. McGuire and L. Tiger, "The Brain and Religious Adaptations," in *Biology of Religious Behavior: The Evolutionary Origins of Faith and Religion*, ed. J. R. Feierman (Santa Barbara, CA: Praeger), pp. 125–40. In the 1600s the German chemist Georg Stahl advanced the idea of a "biomedical soul" that was God given and placed in the body to make things run. See T. S. Hall, *Ideas of Life and Matter* (Chicago: University of Chicago Press, 1969). More recently, the idea of God in the brain has been explored by A. J. Mandell, "Toward a Psychobiology of Transcendence: God in the Brain," in *The Psychobiology of Consciousness*, ed. R. J. Davidson and J. M. Davidson (New York: Plenum, 1980), pp. 379–464.

5. B. Anastas, "The Final Days," *New York Times,* July 1, 2007.

6. A. Rashid, *Taliban* (New Haven, CT: Yale University Press, 2001).

7. H. Smith, *The Religions of Man* (New York: Harper Perennial, 1986).

CHAPTER 2: YOU NEED BOTH A ZOOM LENS AND A MICROSCOPE TO SEE RELIGION

1. "Major Religions Ranked by Size," www.adherents.com (accessed May 14, 2007).

2. "World Religion Statistics," www.adherents.com (accessed June 6, 2009).

3. Worldwide Status of Bible Translation, 2008, www.wycliffe.org, June 7, 2009.

4. D. Henriques and A. Lehren, "Religion for Captive Audiences, with Taxpayers Footing the Bill," *New York Times*, December 10, 2006.

5. "World Religion Statistics."

6. "Atheist Statistics," www.adherents.com, (accessed June 6, 2009).

7. S. Atran and A. Norenzayan, "Religion's Evolutionary Landscape: Counterintuition, Commitment, Compassion, Communion," *Behavioral and Brain Sciences* 27 (2004): 713–70.

8. See the fascinatingly sinuous and elaborate study by T. Cahill, *Mysteries of the Middle Ages: The Rise of Feminism, Science, and Art from the Cults of Catholic Europe* (New York: Doubleday, 2006).

9. B. Graham, *Decision Magazine*, September 2007, p. 4.

10. A. Sullivan, "When Not Seeing Is Believing," *Time*, October 2, 2006.

11. We cannot here attend to the matter, but the broad issue of Marx's and Marxist notions of human nature remains a fascinating one, if rather less practically important now than such notions were way back when. The celebrated prediction about the withering away

of the state under communism has, of course, turned out to be precisely wrong, while the belief in the capacity of intrinsic human macrorationality to organize whole economies and social systems has floundered just as drastically.

12. J. R. Feierman, "How Some Components of Religion Could Have Evolved by Natural Selection," in *The Biological Evolution of Religious Mind and Behavior*, ed. E. Voland and W. Schiefenhovel (New York: Springer-Verlag, 2009), pp. 51–66.

13. W. Z. Zhou et al., "Discrete Hierarchical Organization of Social Group Sizes," *Biological Sciences* 272 (2005): 439–44.

14. C. T. Dawes et al., "Egalitarian Motives in Humans," *Nature* 446 (2007): 794–96.

15. M. Tomasello, *Origins of Human Communication* (Cambridge, MA: MIT Press, 2008).

16. G. Miller, "Probing the Social Mind," *Science* 312 (2006): 838–39.

17. D. C. Dennett, *Breaking the Spell: Religion as a Natural Phenomenon* (New York: Viking, 2005).

18. J. McCrone, "The Power of Belief," *New Scientist*, March 13, 2004.

19. D. C. Dennett, "Show Me the Science," *New York Times*, August 28, 2005.

20. R. Dawkins, *The God Delusion* (Boston: Houghton Mifflin, 2006); S. Harris, *The End of Faith* (New York: Norton, 2004).

21. Sullivan, "When Not Seeing Is Believing."

22. M. B. Norton, *In the Devil's Snare* (New York: Knopf, 2002); J. C. Baroja, *The World of Witches* (London: Phoenix Press, 2001).

23. D. R. Forsyth, "The Function of Attributions," *Social Psychology Quarterly* 43 (1980): 184–89.

24. D. Solomon, "The Nonbeliever," *New York Times Magazine*, January 22, 2006. (This is an interview of Daniel Dennett.)

25. Only Napoleon tried to break that final link in the chain by having all of noble Europe in Notre Dame Cathedral for his coronation as emperor—including the pope. But he ignored the pope and crowned himself. It was a controversial innovation and cleaves to our principle.

26. For example, see the report by Mitchell Krucoff of Duke University Medical School in *Lancet*, July 16, 2005. Prayers from a distance from various religions did not improve the outcome of elective cardiac operations for 737 patients.

27. H. Wilhelm, "The Presbyterian Church Gets into the 9/11 Conspiracy Theory Business," *Wall Street Journal*, September 8, 2006.

28. A. Cooperman and A. S. Tyson, "House Injects Prayer into Defense Bill," *Washington Post*, May 12, 2006.

29. A. Newman, "A Crusade Cannot Thrive by Faith Alone," *New York Times*, June 23, 2005.

30. T. G. Sterling, "Executives Sentenced in Church Fraud," *Washington Post*, October 1, 2006.

31. "Great Faiths: A Journey by Private Jet to the World's Sacred Places," *UCLAlumni Association Magazine* 2006, pp. 1–18.

32. A. Haag, "Church Joins Crusade over Climate Change," *Nature* 449 (2006): 136–37.

33. Quoted by R. Shortt, "The Pope's Divisions," *Times Literary Supplement*, April 10, 2009, p. 23.

CHAPTER 3: ADVENTURES OF THE SOUL

1. G. Ferguson, *Signs and Symbols in Christian Art* (New York: Oxford University Press, 1954).

2. To this day, automobile license plates in Quebec carry the slogan, "*Je me souviens*"—I remember.

3. Handily, the synagogue was on the second floor of the building next to our apartment and visiting it involved but a vault over the fire escape fence we shared.

4. M. Redfern, "Creationist Museum Challenges Evolution," *BBC News*, April 15, 2007.

5. M. J. Behe, *The Edge of Evolution* (New York: Free Press, 2007); M. Holderness, "Enemy at the Gates," *New Scientist*, October 8, 2005; Creation Science Belief Systems, www.religioustolerance.org. March 3, 2005; S. Perkins, "Evolution in Action,"

New Scientist, February 22, 2006; C. Biever, "The God Lab," *New Scientist*, December 16, 2006.

6. N. Bakalar, "Most Doctors See Religion as Beneficial, Study Says," *New York Times*, April 17, 2007.

7. D. MacCulloch, *Reformation: Europe's House Divided, 1490–1700* (London: Allen Lane, 2003).

8. T. L. Thompson, *The Messiah Myth: The Near Eastern Roots of Jesus and David* (London: Cape, 2006).

9. A. Cooperman, "Military Wrestles with Disharmony among Chaplains," *Washington Post*, August 30, 2005. E. J. Dionne Jr., "Keeping Faith with Religious Freedom," *Washington Post*, June 25, 2005.

10. S. Milloy, "PETA: Sacrifice Human, Not Animal Life for Medical Research," *Fox News*, July 25, 2006.

11. R. Dunbar, "We Believe," *New Scientist*, January 28, 2006.

12. C. Soukup, "Religion," *Newsweek*, October 17, 2005.

13. J. Leland, "Sex and the Faithful Soldier," *New York Times*, October 30, 2005.

14. D. MacKenzie, "The End of Enlightenment," *New Scientist*, October 8, 2005.

15. R. E. Nesbett and T. Masuda, "Culture and Point of View," *PNAS* 100, no. 19 (2003): 11163–170; T. E. J. Behrens et al., "Associative Learning of Social Value," *Nature* 456 (2003): 245–49; L. V. Harper, "Epigenetic Inheritance and the Intergenerational Transfer of Experience," *Psychological Bulletin* 131 (2005): 340–60.

CHAPTER 4: FAITH IN SEX

1. J. Joyce, *Portrait of an Artist as a Young Man* (New York: Viking, 1964). *Portrait* was first published in the journal *Egoist* as a series of chapters during 1915–16. The definitive edition, which incorporates Joyce's edits, was published in 1964 by the Viking Press.

2. A. Dante, *Divine Comedy*. The *Comedy* was originally composed between 1310–21 in vernacular Italian. The cited version

is a translation by J. Cardi, *The Divine Comedy (The Inferno, The Purgatorio, and The Paradiso)* (New York: NAL Trade, 2003).

3. All data are from a study of some 35,000 adults conducted by the Pew Forum on Religion and Public Life released on February 25, 2008. It is pertinent that the western states of the United States house the largest proportion of unaffiliated persons, which include atheists and agnostics. It is to be expected that persons who move to nascent communities will be less likely to maintain traditional religious as well as other continuities. It will be important to learn how even larger and speedier patterns of immigration throughout the world will affect the durability of religious commitment. However, according to the study by John Micklehwait and Adrian Wooldridge, the opportunity for choice has produced net growth in religious ranks. This is especially so in the American-style megachurches, which they compare to effective multinational corporations. See J. Micklehwait and A. Wooldridge, *God Is Back: How the Global Revival of Faith Is Changing the World* (New York: Penguin, 2009).

Although we have not addressed the subject in detail, religions frequently are in-group markers for breeding populations. This is one of the reasons why religion is so interested in sex and why some religions say birth control is immoral and other religions (e.g., orthodox Jews and Mormons) encourage their members to have lots of children.

The use of birth control pills among Catholics after having had several children is considered a mortal sin and very likely undermines the teachings and authority of the Church.

4. R. F. Worth, "Challenging Sex Taboos, with Help from the Koran," *New York Times*, June 6, 2009.

5. T. Cairns, *Barbarians, Christians and Muslims* (Cambridge: Cambridge University Press, 1971).

6. An earlier effort by one of the authors to connect religious behavior with religious and ethical constructs was L. Tiger, *The Manufacture of Evil: Ethics, Evolution, and the Industrial System* (New York: Harper & Row, 1987). The central question addressed was, "How does a primate evolved to live in (mainly) hunting-

gathering groups of up to two hundred souls develop the ethical pat-terns to operate in industrial societies of millions of strangers?" The major solution was to adapt the verities of theologians who con-fronted the transition from hunting-gathering to agriculture and pastoralism—of course a daunting ethical crisis. The result was the set of theologies—Islam, Christianity, Judaism, among others—which still, largely, operate in industrial worlds and which reflect that great transformation. Otherwise, "The Lord is my shepherd," for example, is completely baffling. *Per se*, the industrial system struggles still with ethical matters. Marxism, which failed, was one ambitious effort, while utilitarianism suffuses much of economics and law but has produced considerable boredom and no fine archi-tecture, music, and costumes.

CHAPTER 5: RELIGION AS LAW AND THE DENIAL OF BIOLOGY

1. M. Brooks, "In Place of God," *New Scientist*, November 18, 2006.

2. M. Roach, "Almost Human," *National Geographic* 213, no. 4 (2008): 124–45.

3. P. Miller, *The Life of the Mind in America* (New York: Har-court, Brace & World, 1965). The second section of this book, "The Legal Mentality," deals with a question that can be traced back to Blackstone, namely, whether Christianity is part of the law. Jefferson had written a paper that refuted Blackstone. But, having done so, it raised questions about whether lawyers were other than heathens because their practice was without a moral foundation. After all, they could and did readily shift sides.

4. M. Roach, "Almost Human."

5. In the previous chapter we noted that religious affiliations may decline in association with migration, although there were other authors who argue differently; for example, see J. Micklehwait and A. Wooldridge, *God Is Back: How the Global Revival of Faith Is Changing the World* (New York: Penguin, 2009).

6. G. Wallas, *Human Nature in Politics* (New York: Knopf, 1921); W. James, *The Varieties of Religion Experience* (London: Longmans, 1905).

7. E. Voland and W. Schiefenhovel, eds., *The Biological Evolution of Religious Mind and Behavior* (New York: Springer-Verlag, 2009); J. R. Feierman, ed., *The Biology of Religious Behavior: The Evolutionary Origins of Faith and Religion* (Santa Barbara, CA: Praeger, 2009).

8. Africa has an unusually large number of different belief systems, and it is worth considering if many of these developed in different directions due to experiences like those of Chief Fred's village.

9. T. E. J. Behrens et al., "Associative Learning of Social Value," *Nature* 456 (2008): 245–49.

10. L. V. Harper, "Epigenetic Inheritance and the Intergenerational Transfer of Experience," *Psychological Bulletin* 131 (2005): 340–60.

CHAPTER 6: IS RELIGION MONKEY BUSINESS?

1. M. Balter, "Why We're Different: Probing the Gap between Apes and Humans," *Science* 319 (2008): 404–405; R. Orwant, "What Makes Us Human," *New Scientist*, February 21, 2004.

2. C. Floyd, "Virtuous Species: The Biological Origins of Human Morality: An Interview with Frans de Wall," www.sciencespirit.org, March 3, 2005.

3. N. Schultz, "Altruistic Chimpanzees Act for the Benefit of Others," *New Scientist*, June 25, 2007; N. Wade, "Scientist Finds the Beginnings of Morality in Primate Behavior," *New York Times*, March 20, 2007.

4. N. J. Mulcahy and J. Call, "Apes Save Tools for Future Use," *Science* 312 (2006): 1038–40.

5. S. Coulson, "Offerings of a Stone Snake Provide the Earliest Evidence of Religion," *Scientific American*, December 1, 2006.

6. A. Gibbons, "Oldest Members of *Homo sapiens* Discovered in Africa," *Science* 300 (2003): 1641.

7. W. T. Fitch and M. D. Hauser, "Computational Constraints on Syntactic Processing in Nonhuman Primates," *Science* 303 (2004): 377–380

8. F. B. M. de Waal and P. L. Tyack, eds., *Animal Social Complexity* (Cambridge, MA: Harvard University Press, 2004).

9. Note how the acting roles of animals in movies are usually biased strongly toward sharing traits with humans.

10. B. Bower, "Chimpanzees Share Altruistic Capacity with People," *Science News* 171 (2007): 406; G. Vogel, "The Evolution of the Golden Rule," *Science* 303 (2004): 1128–31. In 1971, L. Tiger and R. Fox in *The Imperial Animal* (New York: Holt and Rinehart, 1971) suggested that humans evolved strong ethics for sharing as a consequence of cooperative hunting and gathering. This augmented the basic integration of complex primate societies recently identified in these and other publications.

11. A. Whiten et al., "Conformity to Cultural Norms of Tool Use in Chimpanzees," *Nature* 437 (2005): 737–40; J. Cohen, "The World through a Chimpanzee's Eyes," *Science* 316 (2007): 44–45.

12. T. Breuer, "Gorilla Uses Tool to Plumb the Depths of Abstract Thinking," *New Scientist*, October 8, 2005.

13. J. B. Silk et al., "Chimpanzees Are Indifferent to the Welfare of Unrelated Group Members," *Nature* 437 (2005): 1357–59.

14. M. MacLeod and D. Graham-Rowe, "Every Primate's Guide to Shomoozing," *New Scientist*, September 3, 2005.

15. R. C. Savin-Williams, "An Ethological Study of Dominance Formation and Maintenance in a Group of Human Adolescents," *Child Development* 47 (1976): 972–79.

16. M. Kaplan, "Make Love, Not War," *New Scientist*, December 2, 2006.

17. I. Parker, "Swingers: Bonobos Are Celebrated as Peace-Loving, Matriarchal, and Sexually Liberated. Are They?" *New Yorker*, July 30, 2007.

18. For birds, see S. J. Shettleworth, "Planning for Breakfast," *Nature* 445 (2007): 825–26. For great apes, see T. Suddendori, "Foresight and Evolution of the Human Mind," *Science* 312 (2006): 1006–1007.

19. M. Mesterton-Gibbons and E. S. Adams, "The Economics of Animal Cooperation," *Science* 298 (2002): 2146–47.

20. J. McCrone, "Smarter Than the Average Bug," *New Scientist*, May 21, 2006.

21. B. Holmes, "Did Humans and Chimpanzees Once Merge?" *New Scientist*, May 20, 2006.

22. P. Khaitovich et al., "Parallel Patterns of Evolution in the Genomes and Transcriptomes of Humans and Chimpanzees," *Science* 309 (2005): 1850–54; B. Wood, "Who Are We?" *New Scientist*, October 26, 2002.

23. M. Snyder and M. Gerstein, "Defining Genes in the Genomics Era," *Science* 300 (2003): 258–60; S. E. Ceinlker et al., "Unlocking the Secrets of the Genome," *Nature* 459 (2009): 927–30.

24. "The Chimpanzee Genome," *Nature* 437 (2005): 47–66. Chimpanzee Sequencing and Analysis Consortium, "Initial Sequence of the Chimpanzee Genome and Comparison with the Human Genome," *Nature* 437 (2005): 69–87; International Chimpanzee Chromosome 22 Consortium, "DNA Sequence and Comparative Analysis of Chimpanzee Chromosome 22," *Nature* 429 (2004): 382–88.

25. N. Wade, "Still Evolving, Human Genes Tell New Story," *New York Times*, March 25, 2006.

26. M. Balter, "Brain Evolution Studies Go Micro," *Science* 315 (2007): 1208–11.

27. L. Spinney, "What Only a Chimpanzee Knows," *New Scientist*, June 10, 2006; E. Pennsie, "Nonhuman Primates Demonstrate Humanlike Reasoning," *Science* 317 (2007): 1308.

28. S. Goudarzi, "New Study: The Brain Is Chaotic," www.livescience.com, February 27, 2007.

29. A. S. Deinard and K. K. Kidd, "Evolution of D2 Dopamine Receptor Intron within the Great Apes and Humans," *DNA Sequencing* 8 (1998): 289–301.

30. K. J. Livak et al., "Variability of Dopamine D4 Receptor (DRD4) Gene Sequence within and among Nonhuman Primate Species," *Proceedings National Academy Science USA* 92, no. 2 (1995): 427–31.

31. J. F. Pregenzer et al., "Characterization of Ligand Binding Properties of the 5-HT1D Receptors Cloned from Chimpanzee, Gorilla and Rhesus Monkey in Comparison with Those from the Human and Guinea Pig Receptors," *Neuroscience Letters* 235, no. 3 (1997): 117–20.

32. Unlike in humans, these shifts in mood and behavior are often rapid.

33. V. Morell, "Minds of Their Own," *National Geographic*, March 2008, pp. 37–61; S. Milius, "Ape Aces Memory Tests, Outscores People," *Science News* 172 (2007): 355–56; S. F. Brosman and F. B. M. de Waal, "Monkeys Reject Unequal Pay," *Nature* 425 (2003): 297–99; P. Bloom, "Is God an Accident?" *Atlantic Monthly*, December 2005.

34. P. Bloom, "Children Think before They Speak," *Nature* 430 (2004): 410–11.

35. B. Pesaran et al., "Free Choice Activates a Decision Circuit between Frontal and Parietal Cortex," *Nature* 453 (2008): 406–409.

36. P. Cizek and J. F. Kalaska, "Neural Correlates of Mental Rehearsal in Dorsal Premotor Cortex," *Nature* 431 (2004): 993–96; K. Nelissen et al., "Observing Others: Multiple Action Representation in the Frontal Lobe," *Science* 310 (2005): 332–36.

37. Personal communication. See also S. Pinker, "The Moral Instinct," *New York Times*, January 13, 2008.

38. M. Koenigs et al., "Damage to the Prefrontal Cortex Increases Utilitarian Moral Judgments," *Nature* 446 (2007): 908–11.

39. R. Dunbar, "We Believe," *New Scientist*, January 28, 2006.

40. D. S. Wilson, "Evolution of Religion: The Transformation of the Obvious," in *The Evolution of Religion: Studies, Theories, and Critiques*, ed. J. Bulbulia et al. (Santa Margarita, CA :Collins Foundation Press, 2008), pp. 23–30.

41. See Robert Wright, who has published a book asserting that the notion of universal brotherhood that religions support especially for their own group has had its own evolutionary trajectory resulting efficiently in the notion of God. R. Wright, *The Evolution of God* (Boston: Little, Brown, 2009).

CHAPTER 7: MY BRAIN. YOUR LITURGY.
OUR STATE OF GRACE.

1. Z. Zhou et al., "Genetic Variation in Human *NPY* Expression Affects Stress Response and Emotion," *Nature* 452 (2008): 997–1001.

2. C.-B. Zhong and K. Liljenquist, "Washing Away Your Sins: Threatened Moral and Physical Cleansing," *Science* 313 (2006): 1451–52.

3. See chapter 4, note 1.

4. D. V. Erdman, ed., *The Poetry and Prose of William Blake* (Garden City, NY: Doubleday, 1965), p. 229.

The question of whether God causes religion or religion causes God is discussed by J. Campbell, *The Many Faces of God* (New York: Norton, 2006). On page 216 he states, "The idea that religion produces God, rather than God giving rise to religion, is apt to breed some curious and outlandish offspring. It can issue in the form of a radically decentered faith summed up in the slogan, 'I decide what God is.'"

This is not the idea that we have addressed here. In essence, we are proposing that the brain imagines personages, things, events, scenarios, causalities, and so on, that it cannot completely pin down. In addition, the brain senses uncertainty, ambiguity, and (often) fear—states that it finds aversive—about what it has imagined and cannot prove. It then creates systems to reduce ambiguity and uncertainty. Religion is one of these systems. And it is not irreverent to add double-entry bookkeeping to this category of redemptive sources of reassurance about worldly events.

5. For examples of the use of different methods that lead to the highly similar findings, see S. Hamann and H. Mao, "Positive and Negative Emotional Verbal Stimuli Elicit Activity in the Left Amygdala," *Neuroreport* 13, no. 1 (2002): 15–19; J. J. Paton et al., "The Primate Amygdala Represents the Positive and Negative Value of Visual Stimuli During Learning," *Nature* 439 (2006): 865–70; T. Canli et al., "Amygdala Response to Happy Faces as a Function of Extroversion," *Science* 296 (2002): 2191; L. Helmuth, "Fear and

Trembling in the Amygdala," *Science* 300 (2003): 568–69; R. J. Dolan, "Emotion, Cognition, and Behavior," *Science* 298 (2002): 1191–92.

6. R. B. Adams Jr. et al., "Effects of Gaze on Amygdala Sensitivity to Anger and Fear Faces," *Science* 300 (2003): 1536.

7. P. J. Whalen et al., "Human Amygdala Responsivity to Masked Fearful Eye Whites," *Science* 306 (2004): 2061.

8. T. Singer et al., "Empathy for Pain Involves the Affective but Not the Sensory Components of Pain," *Science* 303 (2004): 1157–62.

9. C. Holden, "Imaging Studies Show How Brain Thinks about Pain," *Science* 303 (2004): 1121.

10. N. I. Eisenberger et al., "Does Rejection Hurt? An fMRI Study of Social Exclusion," *Science* 302 (2003): 290–92; G. MacDonald and M. R. Leary, "Why Does Social Exclusion Hurt? The Relationship between Social and Physical Pain," *Psychological Bulletin* 131 (2005): 202–23.

11. R. L. Trivers, "The Evolution of Reciprocal Altruism," *Quarterly Review of Biology* 46 (1971): 35–57; E. Fehr and U. Fischbacker, "The Nature of Human Altruism," *Nature* 425 (2003): 785–91.

12. T. Singer et al., "Empathetic Neural Responses Are Modulated by the Perceived Fairness of Others," *Nature* 439 (2006): 466–69.

13. R. I. M. Dunbar and S. Shultz, "Evolution in the Social Brain," *Science* 317 (2007): 1344–47.

14. A. Troisi, "Gender Differences in Vulnerability to Social Stress: A Darwinian Perspective," *Physiology and Behavior* 73, no. 3 (2001): 443–49; J. J. Wang et al., "Brain Imaging Shows How Men and Women Cope Differently under Stress," *Social Cognitive and Affective Neuroscience* 24 (2007): 58–61.

15. J. Panksepp, "Affective Consciousness: Core Emotional Feelings in Animals and Humans," *Consciousness and Cognition,* in press.

16. J. N. Wood and J. Grafman, "Human Prefrontal Cortex: Processing and Representational Perspectives," *Nature Reviews* 4 (2003): 139–47.

17. A. G. Hohmann et al., "An Endocannabinoid Mechanism for Stress-Induced Analgesia," *Nature* 435 (2005): 1008–12.

18. G. F. Koob, "Corticotropin-Releasing Factor, Norepinephrine, and Stress," *Biological Psychiatry* 46, no. 9 (1999): 1167–80; B. S. McEwen, "Protective and Damaging Effects of Stress Mediators: Central Role of the Brain," *Dialogues in Clinical Neuroscience* 8, no.4 (2006): 367–81; S. F. Anestis et al., "Age, Rank, and Personality Effects on the Cortisol Sedation Stress Response in Young Chimpanzees," *Physiology and Behavior* 89, no. 2 (2006): 287–94. Studies of chronic stress in mice are consistent with the ideas developed in text. For example, chronic stress leads to a bias in behavioral strategies toward habit. See E. Dias-Ferreira et al., "Chronic Stress Causes Frontostriatal Reorganization and Affects Decision-Making," *Science* 325 (2009): 621–25.

19. C. A. Morilak et al., "Role of Brain Norepinephrine in the Behavioral Response to Stress," *Progressive Neuropsycho-Pharmacology Biological Psychiatry*, October 12, 2005.

20. R. M. Sapolsky, "The Endocrine Stress-Response and Social Status in the Wild Baboon," *Hormones and Behavior* 16 (1982): 279–92; S. F. Anestis et al., "Age, Rank, and Personality Effects on the Cortisol Sedation Stress Response in Young Chimpanzees."

21. H. Pilcher, "The Science of Voodoo: When Mind Attacks Body," *New Scientist*, May 13, 2009.

22. M. Marmot, *The Status Syndrome* (New York: Owl-Books/Henry Holt, 2005).

23. Sapolsky, "The Endocrine Stress-Response and Social Status in the Wild Baboon."

24. A. Mazur and A. Booth, "Testosterone Change after Losing Predicts the Decision to Compete Again," *Hormones and Behavior* 50 (1998) 684–92; J. K. Maner et al., "Submitting to Defeat: Social Anxiety, Dominance Threat, and Decrements in Testosterone," *Psychological Science* 19 (2008): 764–68.

CHAPTER 8: THE ELEPHANT IN THE CHAPEL IS IN YOUR SKULL

1. D. Y. Tsao et al., "A Cortical Region Consisting Entirely of Face-Selective Cells," *Science* 311 (2006): 670–74; F. Formisano et al., "'Who' Is Saying 'What'? Brain-Based Decoding of Human Voice and Speech," *Science* 322 (2008): 970–73; M. S. George et al., "Brain Regions Involved in Recognizing Facial Emotion or Identity: An Oxygen-15 Pet Study," *Journal of Neuropsychiatry* 5 (1993): 384–94.

2. G. Rizzolatti and C. Sinigaglia, *Reflecting on the Mind* (Oxford: Oxford University Press, 2007); G. Miller, "Reflecting on Another's Mind," *Science* 308 (2005): 945–47; C. Zimmer, "How the Mind Reads Other Minds," *Science* 300 (2003): 1079–80.

3. L. Aziz-Zadeh, "Brain's Action Center Is All Talk: Strong Mental Link between Actions and Words," www.sciencedaily.com, September 9, 2006.

4. J. J. Paton et al., "The Primate Amygdala Represents the Positive and Negative Value of Visual Stimuli during Learning," *Nature* 439 (2006): 865–70; T. Sharot et al., "Neural Mechanisms Mediating Optimism Bias," *Nature* 450 (2007): 102–105; L. Tiger, *Optimism: The Biology of Hope* (New York: Simon & Schuster, 1979).

5. N. Abe et al., "Deceiving Others: Distinct Neural Responses of the Prefrontal Cortex and Amygdala in Simple Fabrication and Deception with Social Interactions," *Journal of Cognitive Neuroscience* 19 (2007): 287–95.

6. B. Bower, "Monkey See, Monkey Think," *Science News* 167 (2005): 163–64.

7. T. Singer et al., "Empathetic Neural Responses Are Modulated by the Perceived Fairness of Others," *Nature* 439 (2006): 466–69; S. F. Brosman and F. B. M. de Waal, "Monkeys Reject Unequal Pay," *Nature* 425 (2003): 297–99.

8. For a general review, see R. Masters and M. T. McGuire, *The Neurotransmitter Revolution* (Carbondale: Southern Illinois University Press, 1994). For specific details, see M. T. McGuire and

A. Troisi, "Physiological Regulation-Deregulation and Psychiatric Disorders," *Ethology & Sociobiology* 8 suppl. (1987): 95–125. See M. R. A. Chance, ed., *Social Fabrics of the Mind* (Hove, UK: Erlbaum, 1988) for early intrepid but influential ideas on the effect of events in the environment on the chemistry of the brain.

9. D.S. Moskowitz et al., "Tryptophan, Serotonin and Human Social Behavior," *Advances in Experimental Medicine and Biology* 527 (2003): 215–24.

10. B. Knutson et al., "Selective Alteration of Personality and Social Behavior by Serotonergic Intervention," *American Journal of Psychiatry* 155, no. 3 (1998): 373–79. Note that this is a report about normal subjects. A nontrivial percentage of people who take drugs (usually for depression or anxiety) that increase brain levels of serotonin report that they are less able to experience sexual orgasms compared to the period before taking the drugs. This is not surprising in that altered brain serotonin levels associated with anxiety and depression are only one of the many chemical changes that occur in these states. Thus, elevating serotonin levels alone without changes in other chemical abnormalities is likely to have paradoxical effects. This is to say nothing about the business of providing a serotonogenic basis for feelings of higher status but without the corner office or reserved parking space to make it real.

11. M. J. Crockett et al., "Serotonin Modulates Behavioral Reactions to Unfairness," *Science* 320 (2008): 1739.

12. C. S. Carver et al., "Sertonergic Function; Two-Mode Models of Self-Regulation, and Vulnerability to Depression: What Depression Has in Common with Impulsive Aggression," *Psychological Bulletin* 134 (2008): 912–43.

13. H. F. Clarke et al., "Cognitive Inflexibility after Prefrontal Serotonin Depletion," *Science* 304 (2004): 878–80.

14. S. Nisizawa et al., "Differences between Males and Females in Rates of Serotonin Synthesis in Human Brain," *Proceedings National Academy Science USA* 94 (1997): 5308–13.

15. W. S. Tse and A. J. Bond, "Difference in Serotonergic and Noradrenergic Regulation of Human Social Behaviors," *Psychopharmacology* (Berl) 159 (2002): 216–21.

16. M. R. Roesch and C. R. Olson, "Neuronal Activity Related to Reward Value and Motivation in Primate Frontal Cortex," *Science* 304 (2004): 307–10; G. D. Stuber et al., "Reward-Predictive Cues Enhance Excitatory Synaptic Strength onto Midbrain Dopamine Neurons," *Science* 321 (2008): 1690–92.

17. A. Damasio, "Brain Trust," *Nature* 435 (2005): 571–72; M. Kosfeld et al., "Oxytocin Increases Trust in Humans," *Nature* 435 (2005): 673–76; P. J. Zak, "The Neurobiology of Trust," *Scientific American*, June 2008.

18. T. Canli et al., "Amygdala Response to Happy Faces as a Function of Extroversion," *Science* 296 (2002): 2191.

CHAPTER 9: PUZZLES, ANSWERS, AND MORE PUZZLES

1. J. Borg et al., "The Serotonin System and Spiritual Experiences," *American Journal of Psychiatry* 160 (2003): 1965–69.

2. G. Ferguson, *Signs and Symbols in Christian Art* (New York: Oxford University Press, 1954); L. W. Wagner, *American Life* (Chicago: University of Chicago Press, 1953).

3. J. R. Feierman, "How Some Components of Religion Could Have Evolved by Natural Selection," in *The Biological Evolution of Religious Mind and Behavior*, ed. E. Voland and W. Schiefenhovel (New York: Springer-Verlag, 2009), pp. 51–66; J. R. Feierman, "The Evolutionary History of Religious Behavior," in *The Biology of Religious Behavior*, ed. J. R. Feierman (Santa Barbara, CA: Praeger, 2009), pp. 71–86.

4. P. Ball, *The Essence of Tao* (Royston, Hertfordshire, UK: Eagle Editions), pp. 197–98.

In a personal communication, J. Feierman writes, "It is worth distinguishing between rituals whose rules are passed across generations in DNA and those whose rules are passed across generations by social learning. In some instances, the general blueprint appears to be passed across generations by DNA, such as make-oneself-lower-*or*-smaller-*or*- more vulnerable behavior associated with the non-vocal aspect of petitioning prayer." For the further development of this idea, see J. R. Feierman, "How Some Components of Religion Could Have Evolved by Natural Selection."

5. J. T. Farrow and J. R. Herbert, "Breath Suspension during the Transcendental Meditation Technique," *Psychosomatic Medicine* 44, no. 3 (1982): 133–53.

6. T. Kamei et al., "Decrease in Serum Cortisol during Yoga Exercise Correlated with Alpha-Wave Activation," *Perceptual Motor Skills* 90 (2000): 1027–32.

7. F. Travis and R. K. Wallace, "Autonomic and EEG Patterns during Eye-Closed Rest and Transcendental Meditation (TM) Practice: The Basis for a Neural Model of TM Practice," *Conscious Cognition* 8, no. 3 (1999): 302–18.

8. For overviews, see H. Benson, *The Relaxation Response* (New York: Avon Books, 1975) and D. J. Holmes et al., "Effects of TM and Resting on Physiological and Subjective Arousal," *Journal of Personality and Social Psychology* 44 (1980): 245–52.

9. M. Bujatti and P. Riederer, "Serotonin, Noradrenaline, and Dopamine Metabolites in the Transcendental Meditation Technique," *Journal of Neural Transmission* 39 (1976): 257–67.

10. During the final editing of this book, a book by A. Newberg and M. R. Waldman, *How God Changes Your Brain* (New York: Ballantine, 2009) became available. In this well-written and informative book the authors explore how spiritual beliefs and experience affect changes in the brain. See the following related article, S. W. Lazar et al., "Functional Brain Mapping of the Relaxation Response and Meditation," *Neuroreport* 11, no. 7 (2000): 1581–85.

11. R. Ritskes et al., "MRI Scanning during Zen Meditation: The Picture of Enlightenment," *Constructivism in the Human Sciences* 8 (2003): 85–89.

12. K. Wiech et al., "An fMRI Study Measuring Analgesia Enhanced by Religion as a Belief System," *Pain*, September 5, 2008.

13. M. R. Roesch and C. R. Olson, "Neuronal Activity Related to Reward Value and Motivation in Primate Frontal Cortex," *Science* 304 (2004): 307–10; G. D. Stuber et al., "Reward-Predictive Cues Enhance Excitatory Synaptic Strength onto Midbrain

Dopamine Neurons," *Science* 321 (2008): 1690–92; K. A. Burke et al., "The Role of the Orbitofrontal Cortex in the Pursuit of Happiness and More Specific Rewards," *Nature* 454 (2008): 340–44.

14. P. McNamara, "The Motivational Origins of Religious Practices," *Zygon* 37, no.1 (2002): 143–60.

15. J. J. Paton et al., "The Primate Amygdala Represents the Positive and Negative Value of Visual Stimuli during Learning," *Nature* 439 (2006): 865–70; T. Sharot et al., "Neural Mechanisms Mediating Optimism Bias," *Nature* 450 (2007): 102–105.

16. Marshall McLuhan nailed the difference between symbols and words when he asked us to consider the difference between the Stars and Stripes flag and a similar piece of cloth simply saying: American Flag. M. McLuhan, *Understanding Media* (New York: McGraw-Hill, 1964).

17. H. Phillips, "How Life Shapes the Brainscape," *New Scientist*, November 13, 2005; D. B. Polley et al., "Naturalistic Experience Transforms Sensory Maps in the Adult Cortex of Caged Animals," *Nature* 429 (2004): 67–71; G. H. Mead, *Mind, Self, and Society* (Chicago: University of Chicago Press, 1934).

18. S. B. Hofer et al., "Experience Leaves a Lasting Structural Trace in Cortical Circuits," *Nature* 457 (2009): 313–17.

19. K. J. Flannelly et al., "Beliefs, Mental Health, and Evolutionary Threat Assessment Systems of the Brain," *Journal of Nervous and Mental Disease* 195 (2007): 996–1003; K. J. Flannelly et al., "Beliefs about Life-after-Death, Psychiatric Symptomology, and Cognitive Theories of Psychopathology," *Journal Psychology and Theology* 36, no. 2 (2008): 94–103.

20. For different but overlapping perspectives, see P. Bloom, "Religion Is Natural," *Developmental Science* 10, no. 1 (2007): 147–54; A. Traves, "Religious Experience, and the Brain," in *The Evolution of Religion*, ed. J. Bulbulia et al. (Santa Margarita, CA: Collins Foundation Press, 2008), pp. 211–18; P. Boyer, "Bound to Believe?" *Nature* 455 (2008): 1038–39; H. Whitehouse, "Cognitive Evolution and Religion; Cognitive and Religious Evolution," in *The Evolution of Religion*, ed. J. Bulbulia et al. (Santa Margarita, CA: Collins Foundation Press, 2008), pp. 31–42.

21. J. D. Cohen and G. Aston-Jones, "Decision and Uncertainty," *Nature* 436 (2005): 471.

22. Don't try this at home.

23. J. N. Wood and J. Grafman, "Human Prefrontal Cortex: Processing and Representational Perspectives," *Nature Reviews* 4 (2003): 139–47.

24. M. P. Walker et al., "Dissociable States of Human Memory Consolidation and Reconsolidation," *Nature* 425 (2003): 616–20.

25. L. C. Lee et al., "Independent Cellular Processes for Hippocampal Memory Consolidation and Reconsolidation," *Science* 304 (2004): 839–43; B. E. Depue et al., "Frontal Regions Orchestrate Suppression of Emotional Memories via a Two-Phase Process," *Science* 317 (2007): 215–19.

26. L. Pessoa, "Seeing the World in the Same Way," *Science* 303 (2004): 1617–18.

27. S. C. Pepper, *World Hypotheses* (Berkeley: University of California Press, 1942).

CHAPTER 10: AND WHAT'S YOUR *BRAINSOOTHE* SCORE?

1. Z. Zhou et al., "Genetic Variation in Human *NPY* Expression Affects Stress Response and Emotion," *Nature* 452 (2008): 997–1001.

2. A. Troisi, "Gender Differences in Vulnerability to Social Stress: A Darwinian Perspective," *Physiology and Behavior* 73, no. 3 (2001): 443–49.

3. S. B. Baylin and K. E. Schuebel, "The Epigenomic Era Opens," *Nature* 448 (2007): 548–49; A. Eccleston et al., "Epigenetics," *Nature* 447 (2007): 395–440.

CHAPTER 11: RATHER A BEGINNING, NOT A CONCLUSION

1. Versions of this story are found in D. Bell, *The End of Ideology* (New York: Collier Books, 1961); F. Fukuyama, *Our Posthuman Future* (New York: Farrar, Straus and Giroux, 2002); H. N. Smith, *Virgin Land* (Cambridge, MA: Harvard University Press, 1950).

2. B. Stevens, *Wall Street Journal*, May 12, 2009.

3. R. R. Wisse, *Jews and Power* (New York: Nextbook/Schocken, 2007).

4. A. France, *The Gods Will Have Blood* (New York: Penguin, 1980, first published in 1912). See also, T. Carlyle, *The French Revolution* (New York: George Routledge, 1904).

5. R. M. Gerecht, "The Jihad and the Ballot Box," *New York Times*, June 21, 2009.

6. P. Boyer, "Bound to Believe?" *Nature* 455 (2008): 1038–39.

7. P. Bloom, "Religion Is Natural," *Developmental Science* 10 (2007): 147–51.

8. A. Norenzayan and A. F. Shariff, "The Origin and Evolution of Religious Prosociality," *Science* 322 (2008): 58–62.

9. K. J. Flannelly et al., "Beliefs about Life-after-death, Psychiatric Symptomology, and Cognitive Theories of Psychopathology," *Journal of Psychology and Theology* 36 (2008): 94–103; K. J. Flannelly et al., "Beliefs, Mental Health, and Evolutionary Threat Assessment Systems in the Brain," *Journal of Nervous and Mental Disease* 195, no. 12 (2007): 996–1003; M. McGuire and L. Tiger, "The Brain and Religious Adaptations," in *The Biology of Religious Behavior: The Evolutionary Origins of Religious Behavior*, ed. J. R. Feierman (Santa Barbara, CA: Praeger, 2009), pp. 125–40.

Note that we have not addressed the issues of consciousness and free will. This is largely because these fields are in a rapid state of flux, and, also, we chose not to do so.

10. D. S. Wilson, "Evolution and Religion: Theoretical Formation of the Obvious," in *The Evolution of Religion*, ed. J. Bulbulia et al. (Santa Margarita, CA: Collins Foundation Press, 2008), pp. 23–29.

11. F. de Waal, *Primates and Philosophers: How Morality Evolved* (Princeton, NJ: Princeton University Press, 2007).

12. B. Y. Hayden et al., "Fictive Reward Signals in the Anterior Cingulated Cortex," *Science* 324 (2009): 948–50.

13. E. Gellner, "Nature of Society in Social Anthropology," *Philosophy of Science* 30, no. 3 (July 1963).

INDEX